Springer Theses

Recognizing Outstanding Ph.D. Research

Aims and Scope

The series "Springer Theses" brings together a selection of the very best Ph.D. theses from around the world and across the physical sciences. Nominated and endorsed by two recognized specialists, each published volume has been selected for its scientific excellence and the high impact of its contents for the pertinent field of research. For greater accessibility to non-specialists, the published versions include an extended introduction, as well as a foreword by the student's supervisor explaining the special relevance of the work for the field. As a whole, the series will provide a valuable resource both for newcomers to the research fields described, and for other scientists seeking detailed background information on special questions. Finally, it provides an accredited documentation of the valuable contributions made by today's younger generation of scientists.

Theses are accepted into the series by invited nomination only and must fulfill all of the following criteria

- They must be written in good English.
- The topic should fall within the confines of Chemistry, Physics, Earth Sciences, Engineering and related interdisciplinary fields such as Materials, Nanoscience, Chemical Engineering, Complex Systems and Biophysics.
- The work reported in the thesis must represent a significant scientific advance.
- If the thesis includes previously published material, permission to reproduce this must be gained from the respective copyright holder.
- They must have been examined and passed during the 12 months prior to nomination.
- Each thesis should include a foreword by the supervisor outlining the significance of its content.
- The theses should have a clearly defined structure including an introduction accessible to scientists not expert in that particular field.

More information about this series at http://www.springer.com/series/8790

Iberê Kuntz

Gravitational Theories Beyond General Relativity

Doctoral Thesis accepted by
the University of Sussex, Brighton, UK

Author
Dr. Iberê Kuntz
School of Mathematical and Physical Sciences, Theoretical Particle Physics group
University of Sussex
Brighton, UK

Supervisor
Prof. Dr. Xavier Calmet
School of Mathematical and Physical Sciences, Theoretical Particle Physics group
University of Sussex
Brighton, UK

ISSN 2190-5053 ISSN 2190-5061 (electronic)
Springer Theses
ISBN 978-3-030-21199-8 ISBN 978-3-030-21197-4 (eBook)
https://doi.org/10.1007/978-3-030-21197-4

© Springer Nature Switzerland AG 2019
This work is subject to copyright. All rights are reserved by the Publisher, whether the whole or part of the material is concerned, specifically the rights of translation, reprinting, reuse of illustrations, recitation, broadcasting, reproduction on microfilms or in any other physical way, and transmission or information storage and retrieval, electronic adaptation, computer software, or by similar or dissimilar methodology now known or hereafter developed.
The use of general descriptive names, registered names, trademarks, service marks, etc. in this publication does not imply, even in the absence of a specific statement, that such names are exempt from the relevant protective laws and regulations and therefore free for general use.
The publisher, the authors and the editors are safe to assume that the advice and information in this book are believed to be true and accurate at the date of publication. Neither the publisher nor the authors or the editors give a warranty, expressed or implied, with respect to the material contained herein or for any errors or omissions that may have been made. The publisher remains neutral with regard to jurisdictional claims in published maps and institutional affiliations.

This Springer imprint is published by the registered company Springer Nature Switzerland AG
The registered company address is: Gewerbestrasse 11, 6330 Cham, Switzerland

Supervisor's Foreword

It is a great pleasure to write this foreword for the thesis of Dr. Iberê Oliveira Kuntz de Souza. Iberê's Ph.D. work has been excellent and fully deserved to be published as a Springer Thesis.

His work spanned several topics ranging from models of dark matter to quantum gravity to more speculative ideas on modified gravity. His thesis is based on four papers in common and a single-authored paper. In our first paper, we pointed out that Starobinky's inflation could be induced by quantum effects due to a large non-minimal coupling of the Higgs boson to the Ricci scalar. We showed that the Higgs Starobinsky model provides a solution to issues attached to large Higgs field values in the early universe which in a metastable universe would not be a viable option. We verified explicitly that these large quantum corrections do not destabilize Starobinsky's potential.

The question of vacuum stability was further considered in another paper in collaboration with Prof. Ian Moss. It is well known that in the absence of new physics beyond the Standard Model around 10^{10} GeV, the electroweak vacuum is at best metastable. This represents a major challenge for high-scale inflationary models as, during the early rapid expansion of the universe, it seems difficult to understand how the Higgs vacuum would not decay to the true lower vacuum of the theory with catastrophic consequences if inflation took place at a scale above 10^{10} GeV. We have shown that the non-minimal coupling of the Higgs boson to curvature could solve this problem by generating a direct coupling of the Higgs boson to the inflationary potential thereby stabilizing the electroweak vacuum. For specific values of the Higgs field initial condition and of its non-minimal coupling, inflation can drive the Higgs field to the electroweak vacuum quickly during inflation.

In a third paper, we considered gravitational waves in quantum gravity which was a very timely project as gravitational waves had just been observed by the LIGO and Virgo collaborations.

In our fourth paper, we discussed dark matter in modified theories of gravity and considered how to classify such models. We pointed out that an obvious criterion to classify theories of modified gravity is to identify their gravitational degrees of freedom and their couplings to the metric and the matter sector. Using this simple idea, we were able to show that any theory which depends on the curvature invariants is equivalent to general relativity in the presence of new fields that are gravitationally coupled to the energy–momentum tensor. We have shown that they can be shifted into a new energy–momentum tensor. There is no a priori reason to identify these new fields as gravitational degrees of freedom or matter fields. This leads to an equivalence between dark matter particles gravitationally coupled to the standard model fields and modified gravity theories designed to account for the dark matter phenomenon. Due to this ambiguity, it is impossible to differentiate experimentally between these theories and any attempt of doing so should be classified as a mere interpretation of the same phenomenon.

Finally, I encouraged Iberê to develop his scientific independence and encourage him to develop ideas of his own. He came up with an interesting study of the leading order quantum corrections to the gravitational wave backreaction which he published as a single author.

Iberê has been an excellent Ph.D. student, being able to grasp ideas fast and develop them into interesting results. I am confident that he has what it takes to have a successful academic career. I wish him the very best for his future.

Brighton, UK
May 2019

Prof. Xavier Calmet

Abstract

Despite the success of general relativity in explaining classical gravitational phenomena, several problems at the interface between gravitation and high energy physics remain open to date. The purpose of this thesis is to explore classical and quantum gravity in order to improve our understanding of different aspects of gravity, such as dark matter, gravitational waves and inflation. We focus on the class of higher derivative gravity theories as they naturally arise after the quantization of general relativity in the framework of effective field theory.

The inclusion of higher-order curvature invariants to the action always come in the form of new degrees of freedom. From this perspective, we introduce a new formalism to classify gravitational theories based on their degrees of freedom, and in the light of this classification, we argue that dark matter is no different from modified gravity.

Additional degrees of freedom appearing in the quantum gravitational action also affect the behaviour of gravitational waves. We show that gravitational waves are damped due to quantum degrees of freedom, and we investigate the backreaction of these modes. The implications for gravitational wave events, such as the ones recently observed by the Advanced LIGO collaboration, are also discussed.

The early universe can also be studied in this framework. We show how inflation can be accommodated in this formalism via the generation of the Ricci scalar squared, which is triggered by quantum effects due to the non-minimal coupling of the Higgs boson to gravity, avoiding instability issues associated with Higgs inflation. We argue that the non-minimal coupling of the Higgs to the curvature could also solve the vacuum instability issue by producing a large effective mass for the Higgs, which quickly drives the Higgs field back to the electroweak vacuum during inflation.

Acknowledgements

First and foremost, I would like to thank my wife for her companionship, endless support and, above all, friendship. I am also extremely grateful to my family for all their help and encouragement. I am, and will always be, indebted to them.

I thank everyone that is or once was part of the Theoretical Particle Physics group. Particularly, I would like to thank Xavier Calmet for his supervision and guidance. It was a pleasure to work alongside him.

I thank all my friends for every single pint that we had together. I would never have got this far without them (the pints, I mean). In particular, I want to thank Andrew Bond and Sonali Mohapatra for… well, never mind.

Finally, I would like to thank the National Council for Scientific and Technological Development (CNPq—Brazil) for the financial support.

Contents

Chapter 1
Introduction

1.1 Prelude

Over the course of the past hundred years, general relativity has survived every single experimental test. It was able to explain with high accuracy the anomalous precession of the perihelion of Mercury, which had previously disagreed with the predictions from Newton's gravity. It also correctly predicted the value for the light bending, which was twice the value predicted by the Newtonian theory. Other observations, spanning both classical [1] and modern [2, 3] tests, such as the gravitational redshift, post-Newtonian tests, gravitational lensing, Shapiro time delay, tests of the equivalence principle, strong field tests, cosmological tests, have all favoured general relativity (see [4] for a review). This list goes on and on and, by the time of the writing of this thesis, no experiment has ever measured any deviation from general relativity. In fact, recent observations of gravitational waves by the LIGO collaboration have only reinforced how successful general relativity turns out to be [5].

Given the triumph of general relativity, why should we look into modifying it then? Because, as in any other scientific theory, general relativity has its limitations and it is supposed to be taken seriously only within its domain of validity. As Newtonian physics once faced its own limitations, proving itself useless in relativistic and quantum scales for example, general relativity fails tremendously in certain scales. Of course, given the substantial number of evidence, no one questions the validity of general relativity within its scope, in the same way that no one doubts that Newtonian mechanics can be used to study ballistics. It is thanks to this decoupling of scales that we are able to do physics. This is, in fact, the core of effective field theory and the very reason why we can make progress in science.

Although it is not yet clear at what scale general relativity breaks down, there are possible indications that ask for new physics. The discrepancy between the observed and the theoretical galaxy rotation curves (see Fig. 1.1) [6, 7], for example, cannot be accounted for by either general relativity or the standard model of particle physics, indicating that one of these theories must be incomplete. Dark matter has been postulated as a new type of particle that could account for such discrepancy. Current data

© Springer Nature Switzerland AG 2019

I. Kuntz, *Gravitational Theories Beyond General Relativity*,
Springer Theses, https://doi.org/10.1007/978-3-030-21197-4_1

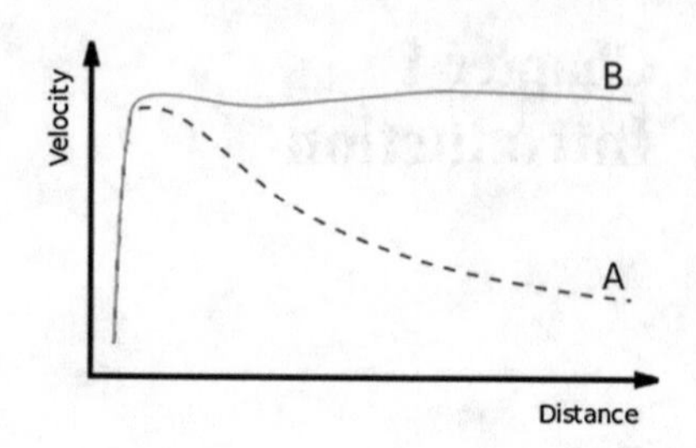

Fig. 1.1 Discrepancy between predicted (A) and observed galaxy rotation curves (B). ©PhilHibbs/Wikimedia Commons/CC-BY-SA-2.0-UK

from the CMB, interpreted in the ΛCDM (Lambda Cold Dark Matter) framework, shows that our universe is made up of approximately 95.1% of an unknown type of energy, where dark matter constitutes 26.8%, dark energy 68.3% and ordinary matter only 4.9% [8]. However, the same observations can be interpreted in a context where general relativity is modified, without the need of postulating new particles. Examples include the tensor-vector-scalar gravity (TeVeS) [9], the scalar-tensor-vector gravity (STVG) [10–12] and $f(R)$ theories [13–15]. TeVeS is a modification of general relativity obtained by the inclusion of new fields to the gravitational sector. It has become popular because it reproduces MOND—a classical modification of Newton's law—in the non-relativistic regime [16–18]. STVG (also known as MOG) was developed via the inclusion of new fields and by promoting some constants of the theory, including the Newton's constant, to scalar fields. As pointed out in [19], the class of theories $f(R)$ where the Einstein-Hilbert action is replaced by a generic function of the Ricci scalar R can also shed new light into the dark matter problem.

Other indications for new physics beyond general relativity also come from late-time cosmology. Dark energy has been hypothesized in order to account for the current acceleration of the universe [20, 21]. The simple addition of a cosmological constant, which is the most economical explanation, leads to other problems, mainly because most quantum field theories predict a cosmological constant that is more than 100 orders of magnitude larger than the measured value [22]. Alternative explanations, such as the inclusion of scalar fields (known as quintessence) [23, 24], are still very popular, but no evidence in its favour has been found so far. Another option would be to modify the gravitational sector in order to explain the accelerated expansion of today's universe.

The inflationary paradigm, initially developed to solve some inconsistencies of the Big Bang cosmology, might also necessitate physics beyond general relativity. In the simplest scenario, a new scalar field dubbed the inflation is required to produce an exponential expansion of the early universe, resulting in the isotropic, homogeneous and flat universe that we observe today [25, 26]. Successful models include the Higgs inflation [27], where the scalar field is described by the Higgs boson, and Starobinsky inflation [28], whose inflaton is hidden in the modification $f(R) = R + R^2$ of general relativity. See Sect. 1.3.1 for a brief review of inflation.

Lastly, there is the problem of quantum gravity, which is perhaps the most challenging problem in theoretical physics. Even though gravity is the oldest of the forces and the only one that is part of everyone's daily lives, it is still the only one lacking a full quantum treatment. Attempts to quantize gravity have led to numerous difficulties over the years, with partial success obtained only in the low-energy regime. While we are still far away from finding the right theory that could describe quantum gravity at, in principle, any energy scale, theoretical advances in the low-energy regime suggests that general relativity must be modified even below the Planck scale [29–31]. The renormalization procedure needed to make quantum general relativity finite at every loop order forces higher-derivative curvature invariants to appear in the action. We will discuss the quantization of general relativity in more detail in Sect. 1.3.2.

In the following sections, we will review basic concepts of general relativity, modified gravity and quantum gravity that will be important in the next chapters. The original contributions start at Chap. 2.

1.2 General Relativity

In this section, we review the geometrical formulation of the general theory of relativity. One postulates that the spacetime is a four-dimensional Pseudo-Riemannian manifold $(\mathcal{M}, g_{\mu\nu})$ composed of a differentiable manifold $\mathcal{M}$ and a metric $g_{\mu\nu}$. Points $p \in \mathcal{M}$ are dubbed events. Test particles, being free from external forces, "free fall" along the spacetime. In a curved manifold, the trajectory of such particles are given by geodesics:

$$\frac{d^2x^\mu}{ds^2} + \Gamma^\mu{}_{\nu\rho}\frac{dx^\nu}{ds}\frac{dx^\rho}{ds} = 0, \tag{1.1}$$

where

$$\Gamma^\rho{}_{\mu\nu} = \frac{g^{\rho\sigma}}{2}\left(\frac{\partial g_{\nu\sigma}}{\partial x^\mu} + \frac{\partial g_{\sigma\mu}}{\partial x^\nu} - \frac{\partial g_{\mu\nu}}{\partial x^\sigma}\right) \tag{1.2}$$

are the Christoffel symbols of the Levi-Civita connection and x^μ are local coordinates. Geodesics followed by massive particles are assumed to be time-like, whereas massless particles, e.g. photons, move along null-like geodesics. Particles that move along space-like geodesics are unphysical as they propagate at superluminal speeds. Such particles are named tachyons. Note that the geodesic equation (1.1) is independent of the particle's mass. This is exactly the equivalence principle: all particles undergo the same acceleration in the presence of a gravitational field, independently of their masses.

The Riemann tensor contains information about the curvature of the spacetime. In coordinates, it is given by

$$R^\rho{}_{\sigma\mu\nu} = \partial_\mu\Gamma^\rho{}_{\nu\sigma} - \partial_\nu\Gamma^\rho{}_{\mu\sigma} + \Gamma^\rho{}_{\mu\lambda}\Gamma^\lambda{}_{\nu\sigma} - \Gamma^\rho{}_{\nu\lambda}\Gamma^\lambda{}_{\mu\sigma}. \tag{1.3}$$

Contracting the first and third indices of the Riemann tensor, one finds the Ricci tensor $R_{\sigma\nu} = g^{\rho\mu} R_{\rho\sigma\mu\nu}$. Contracting the remaining indices of the Ricci tensor, leads to the Ricci scalar $R = g^{\mu\nu} R_{\mu\nu}$.

The dynamics of the gravitational field is described by the Einstein's field equation, which reads

$$R_{\mu\nu} - \frac{1}{2} R g_{\mu\nu} = 8\pi G T_{\mu\nu}, \tag{1.4}$$

where $T_{\mu\nu}$ is the stress-energy tensor of the matter fields. We are using units such that the speed of light is $c = 1$. Equation (1.4) describes the dynamical evolution of the metric $g_{\mu\nu}$, warping and bending spacetime according to the dynamical changes of the matter fields represented by $T_{\mu\nu}$. It is precisely the solutions of (1.4) that have led to the plethora of interesting and successful predictions of general relativity. Observe that Eq. (1.4) cannot be proven from first principles. It was initially obtained by trial and error in an attempt to find a relation between curvature (geometry) and energy (physics).

However, one can adopt a variational approach whose field equations (1.4) could be deduced from. The Einstein-Hilbert action

$$S = \int \mathrm{d}^4 x \sqrt{-g} \frac{1}{16\pi G} R + S_m \tag{1.5}$$

is the most general action containing up to two derivatives of the metric, guaranteeing that the field equation contains up to second orders of the metric. The variation of this action with respect to the metric field leads to (1.4). Needless to say, Eqs. (1.4) and (1.5) are equivalent and have the same physical information. Whether we start from the field equation or from the Lagrangian is just a matter of choice. They offer complementary advantages that can be used accordingly to the problem at hand.

Both the field equation (1.4) and the action (1.5) have an interesting feature. If $\phi : M \to M$ is a diffeomorphism of the spacetime M and $g_{\mu\nu}$ is a solution of (1.4) in the presence of a matter field ψ, then $\phi_* g_{\mu\nu}$ is also a solution of (1.4) in the presence of the matter field $\phi_* \psi$, where ϕ_* denotes the pushforward by ϕ. That is to say that the group of diffeomorphisms is a symmetry group of general relativity in the very same way that $U(1)$ is the symmetry group of electrodynamics. Note that, analogously to gauge theories, the invariance under diffeomorphisms is not a symmetry of the real world as it does not connect two different physical realities to the same description. It is rather a mathematical redundancy that connects two different descriptions to the same physical reality. Therefore, one cannot use such transformations to generate new solutions, but one can exploit this freedom to ease calculations. As we will see in Sect. 1.3.2, however, the importance of the diffeomorphism group is not restricted to easing calculations. It is rather a fundamental principle that guides us on how to look for new physics.

1.2.1 Cosmology

Cosmology is the study of the universe on very large scales. In these scales, one can employ the Copernican principle, which states that the universe is homogeneous (the metric is the same for all points in spacetime) and isotropic (every direction looks the same) on cosmological scales. This is, in fact, what has been observed in the CMB despite very small fluctuations (see Fig. 1.2). The description of a homogeneous and isotropic manifold is given by the Friedmann-Lemaître-Robertson-Walker (FLRW) metric:

$$ds^2 = -dt^2 + a^2(t)\left(\frac{dr^2}{1-kr^2} + r^2(d\theta^2 + \sin^2\theta d\phi^2)\right), \tag{1.6}$$

where $a(t)$ is the scale factor that characterizes the relative size of spacelike hypersurfaces Σ at different times. The curvature parameter k is $+1$ for closed universes, 0 for flat universes and -1 for open universes. In this subsection we will adopt units such that $8\pi G = 1$.

For the ansatz (1.6), the dynamical evolution of the universe is dictated by the scale factor $a(t)$. Its functional form can be found by solving Einstein's equations (1.4) with the input (1.6). Let us assume that the universe is dominated by a perfect fluid with an energy-momentum tensor given by

$$T_{\mu\nu} = (p+\rho)u_\mu u_\nu + p g_{\mu\nu}, \tag{1.7}$$

where $u^\mu = \frac{dx^\mu}{d\tau}$ is the 4-velocity vector field of the fluid, p is the fluid's pressure and ρ is its energy density. Then Einstein's equation for an FLRW metric becomes

Fig. 1.2 All-sky mollweide map of CMB obtained by the WMAP experiment. This image shows a temperature range of $\pm 200\,\mu$K [32]

$$H^2 \equiv \left(\frac{\dot{a}}{a}\right)^2 = \frac{1}{3}\rho - \frac{k}{a^2}, \tag{1.8}$$

$$\dot{H} + H^2 = \frac{\ddot{a}}{a} = -\frac{1}{6}(\rho + 3p), \tag{1.9}$$

where overdots stand for derivative with respect to time t and H is the Hubble parameter. Equations (1.8) and (1.9) are known as Friedmann equations and they describe together the entire structure and evolution of an isotropic and homogeneous universe.

Friedmann equations (1.8) and (1.9) can be combined into the continuity equation

$$\frac{d\rho}{dt} + 3H(\rho + p) = 0, \tag{1.10}$$

which may also be written as

$$\frac{d\ln\rho}{d\ln a} = -3H(1+\omega) \tag{1.11}$$

for the equation of state

$$\omega = \frac{p}{\rho}. \tag{1.12}$$

Integrating Eq. (1.11) and using Eq. (1.8) leads to the solution for the scale factor:

$$a(t) \propto \begin{cases} t^{\frac{2}{3(1+\omega)}}, & \omega \neq -1, \\ e^{Ht}, & \omega = -1. \end{cases} \tag{1.13}$$

This shows that the qualitative behavior of the cosmological evolution depends crucially on the equation of state ω. This fact will be further explored when studying inflation in Sect. 1.3.1, where we will be looking for fluids that violate the strong energy condition $1 + 3\omega > 0$.

1.2.2 *Gravitational Waves*

Gravitational waves are one of the main predictions of general relativity (see e.g. [33] for an extensive review on the subject). They are tiny perturbations of the metric that propagate in spacetime, stretching it and causing observable effects on test particles. The first direct observation was made only in September 2015 by the LIGO collaboration [5] and is considered by many the beginning of a new era in astronomy.

To study gravitational waves, one has to split the metric into a background metric and fluctuations that will be interpreted as the gravitational waves themselves. As a first approximation, we consider gravitational waves propagating in a Minkowski spacetime and we write

$$g_{\mu\nu} = \eta_{\mu\nu} + h_{\mu\nu}. \tag{1.14}$$

Plugging (1.14) into (1.4) leads to

$$\Box \bar{h}_{\mu\nu} + \eta_{\mu\nu}\partial^\rho\partial^\sigma \bar{h}_{\rho\sigma} - \partial^\rho\partial_\nu \bar{h}_{\mu\rho} - \partial^\rho\partial_\mu \bar{h}_{\nu\rho} = 16\pi G T_{\mu\nu}, \tag{1.15}$$

where we have made the field redefinition $\bar{h}_{\mu\nu} = h_{\mu\nu} - \frac{1}{2}\eta_{\mu\nu}h$. We can now use the invariance under diffeomorphisms discussed above to simplify Eq. (1.15). In the linear regime (1.14), a diffeomorphism locally becomes

$$x^\mu \to x'^\mu = x^\mu + \xi^\mu(x). \tag{1.16}$$

Consequently, under (1.16) the field $h_{\mu\nu}$ transforms as

$$h_{\mu\nu}(x) \to h'_{\mu\nu}(x') = h_{\mu\nu}(x) - (\partial_\mu\xi_\nu + \partial_\nu\xi_\mu). \tag{1.17}$$

We can now take advantage of the freedom to choose ξ_μ to simplify (1.15). In fact, one can choose the harmonic gauge

$$\partial^\nu \bar{h}_{\mu\nu} = 0. \tag{1.18}$$

With this choice of gauge, (1.15) becomes

$$\Box \bar{h}_{\mu\nu} = 16\pi G T_{\mu\nu}, \tag{1.19}$$

which is the classical equation of a wave. We conclude that the perturbation of the metric $\bar{h}_{\mu\nu}$ behaves as a wave. Note that, from (1.18) and (1.19), one finds the conservation of the energy-momentum tensor

$$\partial^\nu T_{\mu\nu} = 0. \tag{1.20}$$

Equation (1.20) might seem contradictory as if the energy-momentum conservation holds, then there is no gravitational wave being emitted. This happens because in the linear regime around Minkowski the coupling between gravitational wave and matter is of higher order. It also illustrates that linear gravitational waves cannot carry their own sources, a fact that is also known in electrodynamics where linear eletromagnetic waves are not able to carry electric charges.

To find the energy and momentum carried away by gravitational waves, we must go beyond the linear order in $h_{\mu\nu}$ and figure out the contribution of gravitational waves to the curvature of spacetime. We can no longer use the Minkowski background for this because, otherwise, we would exclude from the beginning the possibility that gravitational waves curve the background. Thus, now we write

$$g_{\mu\nu} = \bar{g}_{\mu\nu} + h_{\mu\nu}, \tag{1.21}$$

where $\bar{g}_{\mu\nu}$ is a dynamical background metric. However, a problem immediately arises as there is no canonical way of defining what part of $g_{\mu\nu}$ is the background and what is the fluctuation. One could, in principle, shift x-dependent terms from $\bar{g}_{\mu\nu}$ to $h_{\mu\nu}$ and vice-versa.

A natural separation between background and fluctuations occurs when there is a clear distinction between their typical scales. Suppose the typical length scale of $\bar{g}_{\mu\nu}$ is its curvature radius L and the length scale of $h_{\mu\nu}$ is its reduced wavelength. If we assume that

$$\frac{\lambda}{L} \ll 1, \tag{1.22}$$

then $h_{\mu\nu}$ has the physical meaning of ripples in the background described by $\bar{g}_{\mu\nu}$. Note that now there are two small parameters: $h = O(|h_{\mu\nu}|)$ and $\epsilon = \lambda/L$. We first expand the equations of motion up to second order in h and then we project out the modes with a short wavelength, i.e. $\epsilon \ll 1$. The simplest way to perform this projection is by averaging over spacetime volume of size d such that $\lambda \ll d \ll L$. In this way, modes with a long wavelength of order L remain unaffected, because they are roughly constant over the volume used for averaging, while modes with a short wavelength of order λ average out because they oscillate very fast.

The separation of gravity into background and fluctuations allows one to expand metric-dependent quantities as

$$R_{\mu\nu} = \bar{R}_{\mu\nu} + R^{(1)}_{\mu\nu} + R^{(2)}_{\mu\nu} + O(h^3), \tag{1.23}$$

where the bar quantities are calculated with respect to the background and the rest depends only on the fluctuation. The superscript (n) is used to indicate the order in h of the underlying term. The resulting Einstein's field equations, after expanding in h and averaging out rapid-oscillating modes, then become

$$\bar{R}_{\mu\nu} - \frac{1}{2}\bar{g}_{\mu\nu}\bar{R} = 8\pi G(\bar{T}_{\mu\nu} + t_{\mu\nu}), \tag{1.24}$$

where

$$t_{\mu\nu} = \frac{-1}{8\pi G}\langle R^{(2)}_{\mu\nu} - \frac{1}{2}\bar{g}_{\mu\nu}R^{(2)}\rangle \tag{1.25}$$

is the energy-momentum contribution from gravitational waves. The brackets in (1.25) denote an average over spacetime, which is responsible for taking only the long-wavelength modes. As it can be seen, the energy and momentum of gravitational waves result from the second order fluctuations of the metric as we had pointed out previously. When the gravitational waves are far away from the source (e.g. at the detector's vicinity), one can further simplify (1.25) by employing the limit of a flat background, imposing the TT gauge

$$h = 0, \quad h_{0\mu} = 0 \tag{1.26}$$

together with the equation of motion $\Box h_{\mu\nu} = 0$. Note that even after choosing the harmonic gauge (1.18), there is still a residual invariance left, which allows us to choose the TT gauge (1.26). In this situation, we find

$$t_{\mu\nu} = \frac{1}{32\pi G}\langle \partial_\mu h_{\alpha\beta}\partial_\nu h^{\alpha\beta}\rangle. \tag{1.27}$$

Observe that, from the covariant conservation of the Einstein tensor

$$\bar{G}_{\mu\nu} = \bar{R}_{\mu\nu} - \tfrac{1}{2}\bar{g}_{\mu\nu}\bar{R} \tag{1.28}$$

with respect to the background connection ∇, one finds from Eq. (1.24) that

$$\nabla^\mu(\bar{T}_{\mu\nu} + t_{\mu\nu}) = 0, \tag{1.29}$$

which shows that there is an exchange of energy and momentum between matter sources and the gravitational waves.

1.3 Modified Gravity

In this section, we review some models of modified gravity that are relevant for this thesis. We start by discussing Lovelock's theorem, which limits the theories one can construct from the metric tensor alone. We then introduce modifications of general relativity in light of Lovelock's result. For a complete review of modified gravity, see [34].

Suppose that the gravitational action contains only the metric field $g_{\mu\nu}$ and its derivatives up to second order. Then, varying the action

$$S = \int \mathrm{d}^4x \mathcal{L}(g_{\mu\nu}) \tag{1.30}$$

leads to the Euler-Lagrange expression

$$E^{\mu\nu} = \frac{d}{dx^\rho}\left[\frac{\partial\mathcal{L}}{\partial g_{\mu\nu,\rho}} - \frac{d}{dx^\lambda}\left(\frac{\partial\mathcal{L}}{\partial g_{\mu\nu,\rho\lambda}}\right)\right] - \frac{\partial\mathcal{L}}{\partial g_{\mu\nu}} \tag{1.31}$$

and the Euler-Lagrange equation $E_{\mu\nu} = 0$. Lovelock's theorem [35, 36] states that the only possible second-order Euler-Lagrange expression obtainable in a four-dimensional space from the action (1.30) is

$$E^{\mu\nu} = \alpha\sqrt{-g}\left[R^{\mu\nu} - \frac{1}{2}g^{\mu\nu}R\right] + \lambda\sqrt{-g}g^{\mu\nu}, \tag{1.32}$$

where α and λ are constants. Therefore, any four-dimensional gravitational action involving only the metric and its derivatives of up to second order leads inevitably to Einstein's equations with or without a cosmological constant.

As a corollary, modifying general relativity requires evading one of the hypotheses of Lovelock's theorem, which includes:

- Considering fields other than the metric;
- Allowing for higher derivatives of the metric;
- Giving up locality;
- Increasing the number of spacetime dimensions;
- Considering other mathematical structures rather than Riemannian manifolds.

In this thesis, we consider the first three of these, focusing mainly on higher derivatives of the metric. As we will see, these three types of modifications are related to each other and they all show up as part of the same formalism.

Let us consider some examples of models that differ from general relativity. The scalar-tensor theories of gravity, whose typical example is Brans-Dicke theory [37], is a modification of general relativity that contains an additional scalar field ϕ coupled to the Ricci scalar:

$$S = \frac{1}{16\pi}\int d^4x\sqrt{-g}\left(\phi R - \frac{\omega(\phi)}{\phi}\partial_\mu\phi\partial^\mu\phi - 2\Lambda(\phi)\right), \tag{1.33}$$

where ω is an arbitrary function and Λ is a ϕ-dependent generalization of the cosmological constant. An important feature of this theory is that under a conformal transformation

$$\tilde{g}_{\mu\nu} = e^{-2\Omega(x)}g_{\mu\nu}, \tag{1.34}$$

where $\Omega(x) = -\frac{1}{2}\ln\phi$, (1.33) can be transformed into general relativity minimally coupled to a scalar field. Performing the transformation (1.34) in the action (1.33) leads to

$$S_E = \int d^4x\sqrt{-\tilde{g}}\left(\frac{1}{16\pi}\tilde{R} - \frac{1}{2}\partial_\mu\psi\partial^\mu\psi - V(\psi)\right), \tag{1.35}$$

where ψ is defined by

$$\frac{\partial\Omega}{\partial\psi} = \sqrt{\frac{4\pi}{3+2\omega}}$$

and

$$V(\psi) = \frac{1}{8\pi} e^{4\Omega} \Lambda$$

is the potential of ψ. Here the objects with the tilde are calculated with the transformed metric $\tilde{g}_{\mu\nu}$. The subscript in S_E stands for Einstein frame, a typical nomenclature used in the literature to refer to the action with the transformed metric $\tilde{g}_{\mu\nu}$, as opposed to the Jordan frame, which refers to the action with the original metric $g_{\mu\nu}$. Therefore, Eq. (1.35) shows that in the Einstein frame the theory becomes the same as general relativity in the presence of the scalar field ψ, which is minimally coupled to gravity through the Jacobian $\sqrt{-\tilde{g}}$. This hidden scalar field is sometimes called scalaron.

There are also theories whose gravitational sector includes other types of fields other than scalars, such as the bimetric theories, tensor-vector-scalar theories (also known as TeVeS) and scalar-tensor-vector theories (not to be confused with TeVeS) [34].

But instead of considering new explicit fields, we can simply consider higher order derivatives in the field equations as opposed to the second order differential equation of general relativity. For example, the class of models described by $f(R)$ [38, 39], i.e.

$$S = \int \mathrm{d}^4x \sqrt{-g} f(R), \tag{1.36}$$

allows for arbitrary powers of the Ricci scalar and, consequently, it produces terms with higher derivatives in the equations of motion. It is important to stress, however, that these theories are equivalent to Brans-Dicke theory (1.33). In fact, let $V(\phi)$ be the Legendre transform of $f(R)$ such that $\phi = f'(R)$ and $R = V'(\phi)$. Then, under a Legendre transformation of (1.36), one obtains the action

$$S = \int \mathrm{d}^4x \sqrt{-g} \left(\phi R - V(\phi)\right), \tag{1.37}$$

which looks exactly like Eq. (1.33) with a potential $V(\phi)$ and $\omega = 0$. By extension, according to (1.35), $f(R)$ is also equivalent to general relativity with a scalar field. This is the nature of the aforementioned relation between additional fields and higher derivative terms. We will see in Chap. 2 that this idea, in fact, extends to more general theories.

An important example of this kind of theory is

$$f(R) = R + \bar{b}_1 R^2, \tag{1.38}$$

known as Starobinsky gravity [28]. This theory successfully explains cosmological inflation by assuming that the inflaton is the scalaron itself. We will see more details of this particular modification in the next subsection.

When considering higher derivatives of the metric, the Ricci scalar is not the only curvature invariant available. Inspired by the renormalization procedure after the quantization of general relativity, other curvatures invariants, such as $R_{\mu\nu}R^{\mu\nu}$ and

$R_{\mu\nu\rho\sigma}R^{\mu\nu\rho\sigma}$, have become equally important [29, 30]. In fact, they are all invariant under the diffeomorphism group and, therefore, should be all considered together. Equation (1.38) then becomes

$$\mathcal{L} = \frac{1}{16\pi G}R + \bar{b}_1 R^2 + \bar{b}_2 R_{\mu\nu}R^{\mu\nu} + \bar{b}_3 R_{\mu\nu\rho\sigma}R^{\mu\nu\rho\sigma}. \tag{1.39}$$

Classically, these terms lead to modifications of the Newton's potential that give rise to Yukawa interactions as shown by Stelle [31]. More importantly, these terms are counterterms that renormalize the quantum gravitational interaction at one-loop order. We will see more on the quantization of the gravitational field in Sect. 1.3.2.

1.3.1 Inflation

Inflation is a period of exponential expansion of the early universe that is believed to have taken place just 10^{-34} s after the Big Bang. First put forward to explain the absence of magnetic monopoles in the universe, inflation later turned out to resolve many other long-standing problems in Big Bang cosmology (see [40] for a review).

The conventional Big Bang theory requires very finely-tuned initial conditions to allow the universe to evolve to its current state. Inflation serves as a bridge between the today's universe and the Big Bang without the need of fine-tuning. Particularly, it explains why the universe we observe is so homogeneous, isotropic and flat.

The comoving particle horizon, i.e. the maximum distance that a light ray can travel between the instants 0 and t, for a universe dominated by a fluid with equation of state $\omega = \frac{p}{\rho}$ is

$$\tau \propto a(t)^{\frac{1}{2}(1+3\omega)}, \tag{1.40}$$

where $a(t)$ is the scale factor of the FLRW universe (1.6). Note that the qualitative behaviour of the comoving horizon τ depends on the sign of $1 + 3\omega$. Fluids satisfying the strong energy condition

$$1 + 3\omega > 0, \tag{1.41}$$

such as matter and radiation dominated universes, would produce a comoving horizon that increases monotonically with time, implying that the regions of the universe entering the horizon today had been far outside the horizon during the CMB decoupling. This leads to the conclusion that causally disjoint patches of the universe yielded a very homogeneous pattern at the CMB, a clear contradiction known as the horizon problem.

Combining Friedmann equation (1.8) with the continuity equation (1.10), one finds

$$\frac{d\Omega}{d\log a} = (1 + 3\omega)\Omega(\Omega - 1), \tag{1.42}$$

where

$$\Omega = \frac{\rho}{\rho_c}. \tag{1.43}$$

The critical energy density $\rho_c = 3H^2$, H being the Hubble constant, is the energy density required for a flat universe. The differential equation (1.42) makes clear that $\Omega = 1$ is an unstable fixed point if the strong energy condition (1.41) is satisfied, thus requiring a finely-tuning initial condition in order to produce a flat universe.

The origin of both the horizon and the flatness problems seem to be related to the strong energy condition. This suggests that a simple solution can be found by violating the relation (1.41), which necessarily requires the fluid pressure to be negative

$$p < -\frac{1}{3}\rho. \tag{1.44}$$

From Friedmann equation (1.9), one can also see that (1.44) is equivalent to an accelerated expansion

$$\frac{d^2a}{dt^2} > 0. \tag{1.45}$$

Equation (1.44) can be satisfied by a nearly constant energy density ρ. The simplest way to do this is by adopting a scalar field—the inflaton—whose potential is sufficiently flat so that the field can slowly roll down the hill (see Fig. 1.3), producing a roughly constant energy density. For this reason, this type of model is known as slow-roll inflation.

To see how this process occurs, let us consider a generic scalar field ϕ minimally coupled to gravity:

$$S = \int \mathrm{d}^4x\sqrt{-g}\left(\frac{1}{16\pi G}R - \frac{1}{2}\partial_\mu\phi\partial^\mu\phi - V(\phi)\right), \tag{1.46}$$

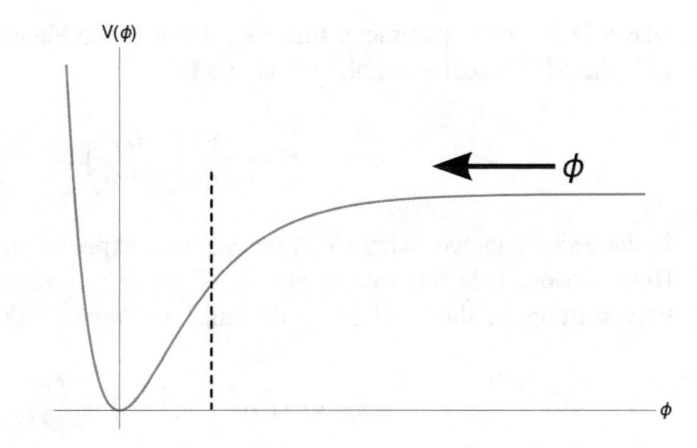

Fig. 1.3 Illustration of slow-roll inflation. The inflaton starts out at the top of the hill and slowly rolls down to smaller values during inflation. The vertical dashed line represents the end of inflation

where $V(\phi)$ denotes the potential of the field ϕ. The energy-momentum tensor for the scalar field is given by

$$T_{\mu\nu} \equiv \frac{-2}{\sqrt{-g}} \frac{\delta S_\phi}{\delta g^{\mu\nu}} = \partial_\mu \phi \partial_\nu \phi - g_{\mu\nu} \left(\frac{1}{2} \partial^\sigma \phi \partial_\sigma \phi + V(\phi) \right), \tag{1.47}$$

where S_ϕ the scalar field action. It follows from (1.47) that the energy density and the pressure of ϕ are given by

$$\rho_\phi = \frac{1}{2}\dot{\phi}^2 + V(\phi), \tag{1.48}$$

$$p_\phi = \frac{1}{2}\dot{\phi}^2 - V(\phi), \tag{1.49}$$

respectively. The resulting equation of state is

$$\omega_\phi = \frac{\frac{1}{2}\dot{\phi}^2 - V(\phi)}{\frac{1}{2}\dot{\phi} + V(\phi)}. \tag{1.50}$$

Therefore, a scalar field ϕ is able to produce inflation if the potential energy $V(\phi)$ dominates over the kinetic energy $\frac{1}{2}\dot{\phi}^2$. In this case, the equation of state becomes $\omega_\phi \approx -1$, which satisfies the condition (1.44).

Now we only need to find a specific description for the scalar field whose potential has the required form described above. Among the sea of models that one can find in the literature, Higgs and Starobinsky inflation stand out as they are both favoured by the CMB constraints [41]. In the former, the inflaton is given by the Higgs field, which is coupled non-minimally to the Ricci scalar [27]:

$$S = \int d^4x \sqrt{-g} \left(\frac{M^2}{2} R + \xi \mathcal{H}^\dagger \mathcal{H} R \right), \tag{1.51}$$

where M is a mass parameter that contributes to the Planck mass $M_p = (8\pi G)^{-1/2}$, $\mathcal{H}$ is the $SU(2)$ scalar doublet which reads

$$\mathcal{H} = \frac{1}{\sqrt{2}} \begin{pmatrix} 0 \\ v + h \end{pmatrix} \tag{1.52}$$

in the unitary gauge, where v is the vacuum expectation value and h denotes the Higgs boson. It is possible to get rid of the non-minimal coupling to gravity by transforming the theory (1.51) to the Einstein frame via the transformation

$$\tilde{g}_{\mu\nu} = \Omega^2 g_{\mu\nu}, \quad \Omega = 1 + \frac{\xi h^2}{M_p^2}. \tag{1.53}$$

This leads to a non-canonical kinetic term for the Higgs field that can be canonically normalized by a field redefinition of the form

$$\frac{d\chi}{dh} = \sqrt{\frac{\Omega^2 + 6\xi^2 h^2/M_p^2}{\Omega^4}}. \tag{1.54}$$

Then, the action in the Einstein frame reads

$$S_E^{\text{Higgs}} = \int d^4x\sqrt{-\tilde{g}}\left(\frac{M_p^2}{2}\tilde{R} - \frac{1}{2}\partial_\mu\chi\partial^\mu\chi - U(\chi)\right), \tag{1.55}$$

where

$$U(\chi) = \frac{1}{\Omega(\chi)^4}\frac{\lambda}{4}\left(h(\chi)^2 - v^2\right)^2. \tag{1.56}$$

For small field values $h \approx \chi$ and $\Omega^2 \approx 1$, thus the potential has the well-known Mexican hat shape of the initial Higgs field h. On the other hand, for large field values of $\chi \gg \sqrt{6}M_p$, one finds [27]

$$h \approx \frac{M_p}{\xi}\exp\frac{\chi}{\sqrt{6}M_p} \tag{1.57}$$

and

$$U(\chi) = \frac{\lambda M_p^4}{4\xi^2}\left(1 + \exp\frac{-2\chi}{\sqrt{6}M_p}\right)^{-2}. \tag{1.58}$$

Hence, the potential $U(\chi)$ is exponentially flat and has the plateau similar to Fig. 1.3, making slow-roll inflation possible.

Starobinsky inflation, on the other hand, is described by the scalaron of Starobinsky gravity (1.38). In the Einstein frame, the theory takes the form [28]

$$S_E^{\text{Starobinsky}} = \int d^4x\sqrt{-\tilde{g}}\left(\frac{1}{16\pi G}\tilde{R} - \frac{1}{2}\partial_{\mu\phi}\partial^\mu\phi - V(\phi)\right), \tag{1.59}$$

where

$$V(\phi) = \frac{M_p^4}{\alpha}\left(1 - \exp\left(-\sqrt{\frac{2}{3}}\frac{\phi}{M_p}\right)\right)^2. \tag{1.60}$$

Thus, it also produces the flatness in the potential for high values of the field ϕ.

1.3.2 *Quantum Gravity*

Although little is known about quantum gravity in the ultraviolet regime, many advances have been achieved in recent years using effective field theory techniques to study low energy quantum gravitational effects [42, 43]. The popular belief that general relativity cannot be quantized is, at best, incomplete and precedes all modern knowledge of quantum field theories. This misconception is commonly associated with the fact that the renormalization procedure generates an infinite number of counterterms in the gravitational action. The coefficient of each counterterm is free and must be fixed by observations, thus indicating that the theory loses its predictive power and becomes unfalsifiable. However, just a small set of free parameters shows up at low energies since high order terms are suppressed by inverse powers of the Planck mass $M_p \sim 10^{19}$ GeV. The high value of M_p is what makes classical general relativity so successful and quantum effects so difficult to probe experimentally.

Divergences appearing at one-loop order, for example, are proportional to R^2, $R_{\mu\nu}R^{\mu\nu}$, $R_{\mu\nu\rho\sigma}R^{\mu\nu\rho\sigma}$, and can be renormalized by the inclusion of counterterms to the Lagrangian [29]:

$$S = \int d^4x\sqrt{-g}\left[\frac{1}{16\pi G}R - \Lambda + \bar{b}_1 R^2 + \bar{b}_2 R_{\mu\nu}R^{\mu\nu} + \bar{b}_3 R_{\mu\nu\rho\sigma}R^{\mu\nu\rho\sigma}\right]. \quad (1.61)$$

The coefficients $\bar{b}_i$ are bare constants and not observables. They are chosen so that divergences at one-loop order turn out to be finite. The curvature squared terms are not all independent due to a topological restriction that occurs only in four dimensions. This relation goes by the name of Gauss-Bonnet theorem and states that the integral of $G = R_{\mu\nu\rho\sigma}R^{\mu\nu\rho\sigma} - 4R_{\mu\nu}R^{\mu\nu} + R^2$ over a compact oriented manifold $\mathcal{M}$ is related to the Euler characteristic $\chi(\mathcal{M})$, thus the integral itself is a topological invariant and its variation results in:

$$\delta\int d^4x\sqrt{-g}\left(R_{\mu\nu\rho\sigma}R^{\mu\nu\rho\sigma} - 4R_{\mu\nu}R^{\mu\nu} + R^2\right) = 0. \quad (1.62)$$

One can then eliminate one of the invariants in terms of the others. We choose to eliminate $R_{\mu\nu\rho\sigma}R^{\mu\nu\rho\sigma}$ and, therefore, we can simply ignore the last term in (1.61).

This theory can be quantized using the background field method [44]. We perturb the metric $g_{\mu\nu} \to g_{\mu\nu} + h_{\mu\nu}$ and integrate out the fluctuations $h_{\mu\nu}$ using the Feynman path integral formalism:

$$e^{-\Gamma} = \int \mathcal{D}h_{\mu\nu}\mathcal{D}\Phi\, e^{-(S[g+h]+S_m[\Phi])}, \quad (1.63)$$

where S_m is the action of matter sector and Φ represents a set of arbitrary matter fields (not necessarily scalar fields). The quantum effective action Γ describes quantum gravitational phenomena and can be used to investigate the phenomenology of quantum gravity at low energies (below the Planck scale). As expected, the general result is quite cumbersome even at the leading order, containing several terms that contribute equally [45]. However, if one considers only the limit of massless or very light fields, the outcome turn out to be very neat. In this limit, non-localities are expected to show up as massless fields mediate long-range interactions. In fact, the quantum action in this case is given by [44, 46–48]

$$\Gamma = \Gamma_L + \Gamma_{NL}, \tag{1.64}$$

where the local part reads

$$\Gamma_L = \int \mathrm{d}^4x\sqrt{-g}\left(\frac{R}{16\pi G} - \Lambda + b_1(\mu)R^2 + b_2(\mu)R_{\mu\nu}R^{\mu\nu}\right), \tag{1.65}$$

and the non-local one reads

$$-\Gamma_{NL} = \int \mathrm{d}^4x\sqrt{-g}\left[c_1 R\ln\left(-\frac{\Box}{\mu^2}\right)R + c_2 R_{\mu\nu}\ln\left(-\frac{\Box}{\mu^2}\right)R^{\mu\nu}\right. \tag{1.66}$$

$$\left. + c_3 R_{\mu\nu\rho\sigma}\ln\left(-\frac{\Box}{\mu^2}\right)R^{\mu\nu\rho\sigma}\right]. \tag{1.67}$$

The log operator is defined as

$$\ln\frac{-\Box}{\mu^2} = \int_0^\infty \mathrm{d}s\left(\frac{1}{\mu^2+s} - G(x,x',\sqrt{s})\right), \tag{1.68}$$

where $G(x,x';\sqrt{s})$ is the Green's function of

$$(-\Box + k^2)G(x,x';k) = \delta^4(x-x') \tag{1.69}$$

with proper boundary conditions. The non-local piece represents the infrared portion of quantum gravity and, as such, it is completely independent of the UV completion. In fact, the coefficients c_i are genuine predictions of the quantum theory of gravity. They are determined once the matter fields Φ that are integrated out in Eq. (1.63) and their respective spins are specified; see Table 1.1. The total contribution to each coefficient is given by simply summing the contribution from each matter species. For example, for N_s minimally coupled scalars ($\xi = 0$) and N_f fermions, we have

$$c_1 = \frac{5}{11520\pi^2}N_s - \frac{5}{11520\pi^2}N_f. \tag{1.70}$$

The local action, on the other hand, represents the high energy portion of quantum gravity. As a result, the coefficients b_i cannot be determined from first principles. They are renormalized parameters which must be fixed by observations (as opposed to $\bar{b}_i$ that are not observables). They depend on the renormalization scale μ in such a way that they cancel the μ-dependence of the non-local logarithm operator. Thus,

Table 1.1 Values of the coefficients c_i for each spin (ξ is the non-minimal coupling coefficient of scalars to gravity) [48]. Each value must be multiplied by the number of fields of its category. The total value of each coefficient is given by adding up all contributions. See Eq. (1.70) for an example

	c_1	c_2	c_3
Real scalar	$5(6\xi - 1)^2/(11520\pi^2)$	$-2/(11520\pi^2)$	$2/(11520\pi^2)$
Dirac spinor	$-5/(11520\pi^2)$	$8/(11520\pi^2)$	$7/(11520\pi^2)$
Vector	$-50/(11520\pi^2)$	$176/(11520\pi^2)$	$-26/(11520\pi^2)$
Graviton	$430/(11520\pi^2)$	$-1444/(11520\pi^2)$	$424/(11520\pi^2)$

the total effective action Γ is independent of μ. The renormalization group equation is

$$\mu \partial_\mu b_i = \beta_i, \tag{1.71}$$

where $\beta_i = -2c_i$ are the beta functions, thus the running of b_i can also be obtained straightforwardly from Table 1.1. The relation between the beta functions of b_i and the coefficients c_i is expected because the resultant action Γ must be independent of μ as explained above.

While we are yet far away from being able to probe quantum gravity experimentally, the above shows that we can use standard techniques from quantum field theory to quantize general relativity. Needless to say, this is the most conservative approach of all. If we insist that general relativity and quantum field theory are correct descriptions of our world below the Planck scale—and as far as observations go, this is indeed true—, we can then quantize general relativity in the very same way that the other interactions are quantized. As a result, we can make genuine and model-independent predictions of quantum gravity without appealing to *ad hoc* hypotheses.

1.4 Outline of This Thesis

This thesis contains a collection of published work that was completed as part of my doctoral degree, which concerns modifications of gravity and its implications to gravitational waves, inflation and dark matter. It is organized as follows:

- In Chap. 2, based on [49], we set up a new formalism to classify gravitational theories based on their degrees of freedom and how they interact with the matter sector. We argue that every modification of the action performed by the inclusion of additional curvature invariants inevitably leads to new degrees of freedom. This can be seen by diagonalizing the action, either via field redefinitions or through the linearization process around a given background, and further canonically normalizing it. A particular example of this is the well-known equivalence of $f(R)$ and general relativity minimally coupled to a scalar field. We also give less obvious examples where invariants such as $R_{\mu\nu}R^{\mu\nu}$ and $R_{\mu\nu\rho\sigma}R^{\mu\nu\rho\sigma}$ are also present. As an application, we consider the dark matter problem and we show that particle dark matter models and modified gravity models are actually equivalent as they are both based on new degrees of freedom.

- Chapter 3 is based on [50] and we study gravitational waves from the effective field theory perspective. We show that one-loop quantum corrections lead to modifications of the analytic structure of the graviton propagator, yielding the so-called dressed propagator for the graviton. The dressed propagator contains additional complex poles, thus effectively leading to new propagating degrees of freedom. The real part is interpreted as the mass of the modes and the imaginary part is interpreted as their width. We study the consequences of these additional degrees of freedom for gravitational waves. Particularly, we show that gravitational waves become damped when quantum gravitational effects are taken into account. The consequences for gravitational wave events, such as GW 150914, recently observed by the Advanced LIGO collaboration, are discussed.
- The effect on the energy of gravitational waves due to one-loop corrections are studied in Chap. 4, which is based on [51]. By performing the short-wave formalism, we separate the modes with a long wavelength from the ones with a short wavelength. The former contains information about the contribution of gravitational waves to the spacetime curvature, while the latter affects the propagation of gravitational waves in curved spacetimes. The energy-momentum tensor $t_{\mu\nu}$ of gravitational waves is then calculated, thus showing how quantum effects contribute to the backreaction of gravitational waves. The trace of the effective energy-momentum tensor is shown to be non-vanishing and, hence, it contributes to the cosmological constant. The first bound on the amplitude of the massive mode is found by comparing the gravitational wave energy density $\rho = t_{00}$ with LIGO's data. In addition, we show that the propagation of gravitational waves in curved spacetimes can be obtained by covariantization of the gravitational wave equation in flat spacetime, i.e. by simply replacing $\eta_{\mu\nu}$ and ∂_μ by $\bar{g}_{\mu\nu}$ and ∇_μ, respectively.
- Chapter 5, which is based on [52], is designated to investigate an interesting interplay between Higgs and Starobinsky inflation. We show that Starobinsky inflation, based on the modification $f(R) = R + \alpha R^2$ of general relativity, can be generated by quantum effects due to the non-minimal coupling of the Higgs to gravity. After quantization of the gravitational action, the coefficient α acquires a dependence on the coefficient ξ of the coupling between the Higgs and the Ricci scalar. For large values of ξ, one obtains the required value for α so that Starobinsky inflation can take place. This formalism avoids instability issues caused by large values of the Higgs boson as the scalaron in the Starobinsky model is the only field required to take large values in the early universe.
- In Chap. 6 [53], we study the instability problem in a more general setting, i.e. when the inflaton is not restricted to be the Higgs field. In these cases, even though inflation is not driven by the Higgs, the direct coupling between the Higgs and the curvature could still cause problems during and after inflation as claimed in [54, 55]. We argue that, after canonically normalizing the Higgs field, an interaction between the inflationary potential and the Higgs is induced. This interaction

produces a large effective mass for the Higgs, which quickly drives the Higgs boson back to the electroweak vacuum during inflation, thus stabilizing the false vacuum.

- Lastly, we draw the conclusions and discuss future directions in Chap. 7.

References

1. Einstein A (1916) The foundation of the general theory of relativity. Ann Phys 49(7):769–822 (65, 1916)
2. Dicke RH (1959) New research on old gravitation. Science 129(3349):621–624
3. Schiff LI (1960) On experimental tests of the general theory of relativity. Am J Phys 28:340–343
4. Will CM (2014) The confrontation between general relativity and experiment. Living Rev Rel 17:4
5. Abbott BP et al (2016) Observation of gravitational waves from a binary black hole merger. Phys Rev Lett 116(6):061102
6. Rubin VC, Ford WK Jr (1970) Rotation of the Andromeda nebula from a spectroscopic survey of emission regions. Astrophys J 159:379–403
7. Rubin VC, Thonnard N, Ford WK Jr (1980) Rotational properties of 21 SC galaxies with a large range of luminosities and radii, from NGC 4605/R = 4kpc/to UGC 2885/R = 122 kpc. Astrophys J 238:471
8. Ade PAR et al (2016) Planck 2015 results XIII. Cosmological parameters. Astron Astrophys 594:A13
9. Bekenstein JD (2004) Relativistic gravitation theory for the MOND paradigm. Phys Rev D 70:083509 (Erratum: Phys Rev D 71:069901, 2005)
10. Moffat JW (2006) Scalar-tensor-vector gravity theory. JCAP 0603:004
11. Brownstein JR, Moffat JW (2006) Galaxy rotation curves without non-baryonic dark matter. Astrophys J 636:721–741
12. Brownstein JR, Moffat JW (2006) Galaxy cluster masses without non-baryonic dark matter. Mon Not R Astron Soc 367:527–540
13. Buchdahl HA (1970) Non-linear Lagrangians and cosmological theory. Mon Not R Astron Soc 150:1
14. Capozziello S, Cardone VF, Carloni S, Troisi A (2004) Can higher order curvature theories explain rotation curves of galaxies? Phys Lett A 326:292–296
15. Katsuragawa T, Matsuzaki S (2017) Dark matter in modified gravity? Phys Rev D 95(4):044040
16. Milgrom M (1983) A modification of the Newtonian dynamics as a possible alternative to the hidden mass hypothesis. Astrophys J 270:365–370
17. Milgrom M (1983) A modification of the Newtonian dynamics: implications for galaxies. Astrophys J 270:371–383
18. Milgrom M (1983) A modification of the Newtonian dynamics: implications for galaxy systems. Astrophys J 270:384–389
19. Capozziello S, De Laurentis M (2012) The dark matter problem from f(R) gravity viewpoint. Ann Phys 524:545–578
20. Peebles PJE, Ratra B (2003) The cosmological constant and dark energy. Rev Mod Phys 75:559–606 (592, 2002)
21. Carroll SM (2001) The cosmological constant. Living Rev Rel 4:1
22. Adler RJ, Casey B, Jacob OC (1995) Vacuum catastrophe: an elementary exposition of the cosmological constant problem. Am J Phys 63:620–626
23. Ratra B, Peebles PJE (1988) Cosmological consequences of a rolling homogeneous scalar field. Phys Rev D 37:3406

24. Caldwell RR, Dave R, Steinhardt PJ (1998) Cosmological imprint of an energy component with general equation of state. Phys Rev Lett 80:1582–1585
25. Linde AD (1982) A new inflationary universe scenario: a possible solution of the horizon, flatness, homogeneity, isotropy and primordial monopole problems. Phys Lett B 108:389–393
26. Albrecht A, Steinhardt PJ (1982) Cosmology for grand unified theories with radiatively induced symmetry breaking. Phys Rev Lett 48:1220–1223
27. Bezrukov FL, Shaposhnikov M (2008) The standard model Higgs boson as the inflaton. Phys Lett B 659:703–706
28. Starobinsky AA (1980) A new type of isotropic cosmological models without singularity. Phys Lett B 91:99–102 (771, 1980)
29. 't Hooft G, Veltman MJG (1974) One loop divergencies in the theory of gravitation. Ann Inst H Poincare Phys Theor A20:69–94
30. Stelle KS (1977) Renormalization of higher derivative quantum gravity. Phys Rev D 16:953–969
31. Stelle KS (1978) Classical gravity with higher derivatives. Gen Rel Grav 9:353–371
32. Bennett C et al (2003) First year Wilkinson microwave anisotropy probe (WMAP) observations: foreground emission. Astrophys J Suppl 148:97
33. Maggiore M (2007) Gravitational waves, vol 1: theory and experiments. Oxford master series in physics. Oxford University Press
34. Clifton T, Ferreira PG, Padilla A, Skordis C (2012) Modified gravity and cosmology. Phys Rept 513:1–189
35. Lovelock D (1971) The Einstein tensor and its generalizations. J Math Phys 12:498–501
36. Lovelock D (1972) The four-dimensionality of space and the Einstein tensor. J Math Phys 13:874–876
37. Brans C, Dicke RH (1961) Mach's principle and a relativistic theory of gravitation. Phys Rev 124:925–935 (142, 1961)
38. Sotiriou TP, Faraoni V (2010) f(R) theories of gravity. Rev Mod Phys 82:451–497
39. De Felice A, Tsujikawa S (2010) f(R) theories. Living Rev Rel 13:3
40. Baumann D (2011) Inflation. Physics of the large and the small, TASI 09. In: Proceedings of the theoretical advanced study institute in elementary particle physics, Boulder, Colorado, USA, 1–26 June 2009, pp 523–686
41. Akrami Y et al (2018) Planck 2018 results. X. Constraints on inflation
42. Buchbinder IL, Odintsov SD, Shapiro IL (1992) Effective action in quantum gravity
43. Vilkovisky GA (1992) Effective action in quantum gravity. Class Quant Grav 9:895–903
44. Barvinsky AO, Vilkovisky GA (1985) The generalized Schwinger-Dewitt technique in gauge theories and quantum gravity. Phys Rept 119:1–74
45. Codello A, Jain RK (2016) On the covariant formalism of the effective field theory of gravity and leading order corrections. Class Quant Grav 33(22):225006
46. Barvinsky AO, Vilkovisky GA (1987) Beyond the Schwinger-Dewitt technique: converting loops into trees and in-in currents. Nucl Phys B 282:163–188
47. Barvinsky AO, Vilkovisky GA (1990) Covariant perturbation theory. 2: second order in the curvature. General algorithms. Nucl Phys B 333:471–511
48. Donoghue JF, El-Menoufi BK (2014) Nonlocal quantum effects in cosmology: quantum memory, nonlocal FLRW equations, and singularity avoidance. Phys Rev D 89(10):104062
49. Calmet X, Kuntz I (2017) What is modified gravity and how to differentiate it from particle dark matter? Eur Phys J C 77(2):132
50. Calmet X, Kuntz I, Mohapatra S (2016) Gravitational waves in effective quantum gravity. Eur Phys J C 76(8):425
51. Kuntz I (2018) Quantum corrections to the gravitational backreaction. Eur Phys J C 78(1):3
52. Calmet X, Kuntz I (2016) Higgs Starobinsky inflation. Eur Phys J C 76(5):289
53. Calmet X, Kuntz I, Moss IG (2018) Non-minimal coupling of the Higgs boson to curvature in an inflationary universe. Found Phys 48(1):110–120
54. Herranen M, Markkanen T, Nurmi S, Rajantie A (2014) Spacetime curvature and the Higgs stability during inflation. Phys Rev Lett 113(21):211102
55. Herranen M, Markkanen T, Nurmi S, Rajantie A (2015) Spacetime curvature and Higgs stability after inflation. Phys Rev Lett 115:241301

Chapter 2
Equivalence Between Dark Matter and Modified Gravity

A useful touchstone to classify theories of modified gravity is to identify their gravitational degrees of freedom together with their couplings to themselves and to the matter sector. By implementing this idea, we show that any theory which depends on the curvature invariants is equivalent to general relativity in the presence of new fields that are gravitationally coupled to the energy-momentum tensor. We demonstrate that the degrees of freedom originated from curvature invariants can be shifted into a new energy-momentum tensor. There is no a priori reason to identify these new fields as gravitational degrees of freedom or matter fields. This leads to an equivalence between dark matter particles coupled gravitationally to the standard model fields and modified gravity theories developed to account for the dark matter phenomenon. Due to this ambiguity, it is hardly possible to distinguish experimentally between these theories and any attempt of doing so should be seen as a mere interpretation of the same phenomenon.

2.1 Introduction

General relativity and the standard model of particle physics have both been extraordinarily successful in describing our universe both on cosmological scales as well as on microscopic scales. Despite this incredible triumph, some observations cannot be explained by these otherwise extremely successful models. For example, the cosmic microwave background, the rotation curves of galaxies or the bullet cluster to quote a few [1], suggest that there is a new form of matter that does not shine in the electromagnetic spectrum. Dark matter is not accounted for by either general

© Springer Nature Switzerland AG 2019

I. Kuntz, *Gravitational Theories Beyond General Relativity*,
Springer Theses, https://doi.org/10.1007/978-3-030-21197-4_2

relativity or the standard model of particle physics.[1] While a large portion of the high energy physics community is convinced that dark matter should be characterized by yet unfound new particles, it remains an open question whether this phenomenon requires a modification of the standard model or of general relativity. Here we would like to raise a slightly different question, namely whether the distinction between modified gravity or new particles is always unambiguous. We will show that the answer to this question turns out to be negative.

Models of modified gravity are attractive given the frustrating success of the standard model at surviving its confrontation with the data of the Large Hadron Collider. Modified gravity theories have been designed in the hope of discovering solutions to the dark matter or dark energy questions. All kinds of theories have been proposed in order to address these issues. Among them, we can find higher derivative gravity theories (e.g. $f(R)$), the scalar-tensor theories (e.g. Brans-Dicke), the non-metric theories (e.g. Einstein-Cartan theory), just to cite a few, see [4] for a solid review.

In the context of quantum field theories, fields are just dummy variables as the action is formulated as a path integral over all field configurations. This entails a reparametrization invariance of field theories. In gravitational theories (see e.g. [5]), this corresponds simply to the freedom to pick a specific frame to define one's model. The reparametrization invariance makes it difficult to distinguish between the plethora of theories as depending on which field variables are chosen, the very same model could seem to be very different in two distinct frames. One of the purposes of this chapter is to apply a very simple and obvious criterion to classify gravitational theories. The idea is to identify their gravitational degrees of freedom by looking at the poles in the field equations and carefully identifying the coupling of these poles to the metric and the energy-momentum tensor (matter sector). This allows one to unambiguously compare two gravitational theories. Some work in this direction was done in the past [6], but the focus was given to the different action principles, namely the metric, metric-affine and affine formalisms. Here we present a broader approach which can be applied to any kind of theory independently of its action principle.

In this chapter, we aim to propose a general framework where gravitational theories can be compared to each other in order to classify them into different classes of physically equivalent theories. The classification method will be presented in Sect. 2.2 together with some examples. In Sect. 2.3 we apply these ideas to the dark matter problem and show that the distinction between modified gravity or dark matter as a new particle is not always so clear. In particular, we show that any theory which depends on the curvature invariants is equivalent to general relativity in the presence of new fields that are gravitationally coupled to the energy-momentum tensor. We show that they can be shifted into a new energy-momentum tensor. Modified dark matter is thus equivalent to new degrees of freedom (i.e. particles) that are coupled gravitationally to regular matter. We then make the conclusions in Sect. 2.4.

---

[1] One should note though that the possibility of Planck mass quantum black holes remnants [2, 3] is not excluded, but it is difficult to find an inflationary model that produces them at the end of inflation.

2.2 Classification of Extended Theories of Gravity

Fields in a quantum field theory are dummy variables, i.e. they only appear in intermediate steps as integration variables. The metric in a gravitational theory plays the role of a field and thus is subject to the same kind of invariance under metric redefinitions. Therefore two apparently very different gravitational theories can actually turn out to be mathematically equivalent when expressed using different "coordinates" in field space. An important example is given by $f(R)$ theories:

$$S = \int d^4x\sqrt{-g}\left(\frac{1}{16\pi G}f(R) + \mathcal{L}_M\right) \tag{2.1}$$

where $f(R)$ is an arbitrary function of the Ricci scalar. When transforming the action from the Jordan to the Einstein frame it becomes clear that $f(R)$ is equivalent to general relativity in the presence of a scalar field that is gravitationally coupled to the matter sector. In fact, it is well known that after a Legendre transformation followed by a conformal rescaling $\tilde{g}_{\mu\nu} = f'(R)g_{\mu\nu}$, $f(R)$ theory can be put in the form [7]

$$\begin{aligned} S = &\int d^4x\sqrt{-\tilde{g}}\left(\frac{1}{16\pi G}\tilde{R} - \frac{1}{2}\tilde{g}^{\mu\nu}\partial_\mu\phi\partial_\nu\phi - V(\phi)\right) \\ &+ \int d^4x\sqrt{-\tilde{g}}F^{-2}(\phi)\mathcal{L}_M(F^{-1}(\phi)\tilde{g}_{\mu\nu}, \psi_M), \end{aligned} \tag{2.2}$$

where

$$\phi \equiv \sqrt{\frac{3}{16\pi G}}\log F, \tag{2.3}$$

$$F(\phi) \equiv f'(R(\phi)). \tag{2.4}$$

Thus all matter fields acquire a universal coupling to a new scalar field ϕ through the factor $F^{-1}(\phi)$. Massless gauge bosons are exceptions since their Lagrangians are invariant under conformal transformations of the metric. This simple example shows that, despite the apparent simplicity of $f(R)$ which naively seems to only depend on the metric $g_{\mu\nu}$, the theory also contains an extra scalar degree of freedom. The reason for the presence of this additional degree of freedom can be understood using group representation theory. A particle is defined as an irreducible representation of the Poincaré group and it is fully characterized by its mass and spin. Irreducible representations are then embedded into fields that are used in the definition of a Lagrangian of certain particle of a given mass and spin. For example, any symmetric tensor field $g_{\mu\nu}$, including the metric, can be decomposed into its irreducible spin representations:

$$g_{\mu\nu} \in \mathbf{0} \oplus \mathbf{1} \oplus \mathbf{2} \tag{2.5}$$

and we see that there are three potential particle representations in a symmetric tensor field. The Einstein-Hilbert action takes care of killing the dynamics of the spin-0 and spin-1 degrees of freedom, only exciting the spin-2 particle, i.e. the usual graviton. However, when one modifies the gravitational action, the other degrees of freedom might become dynamical too. It must be stressed that the degrees of freedom are the same in all frames. When we map the theory to the Einstein frame, we are just picking up variables so to make the extra degrees of freedom explicit in the action. But be advised, the additional degrees of freedom are by no means an artefact of the transformation!

The above example can be generalized to any gravitational theory. A generic gravitational theory, assuming that it is a metric theory, will have at least one metric tensor (if it is to have general relativity in some limit) and other fields of different spins. We assume that this theory can be described by an action $S = S[\phi^1_{\alpha_1}, \ldots, \phi^n_{\alpha_n}]$, where $\phi^i_{\alpha_i}$ are the fields and α_i represents generically the number of indices, i.e. the type of the field (e.g. scalar, tensor, etc). The coupling of the gravitational degrees of freedom to matter $\mathcal{L}_{\mathcal{M}}$ needs to be specified. An algorithm to classify gravitational theories, in the sense of comparing two gravitational theories, can be designed as follows.

(1) The first step is to find all gravitational degrees of freedom of each theory.
(2) Verify how these degrees of freedom couple to the matter fields as well as to themselves.

The first step might seem straightforward if we have in mind theories with canonical Lagrangians. Nonetheless, this is not the case for gravitational theories where degrees of freedom are hidden in the higher derivative terms appearing in the action. The identification of degrees of freedom can be performed by linearizing the equations of motion around some fixed background $g_{\mu\nu} = g^{(0)}_{\mu\nu} + h_{\mu\nu}$ and looking for the poles of the propagator $\mathcal{P}_{\alpha\beta\mu\nu}$:

$$\mathcal{D}_{\alpha\beta\mu\nu}h^{\mu\nu} = T_{\alpha\beta} \implies \mathcal{P}_{\alpha\beta\mu\nu} = \mathcal{D}^{-1}_{\alpha\beta\mu\nu}, \tag{2.6}$$

where $\mathcal{D}_{\alpha\beta\mu\nu}$ is the modified wave operator. The position of the poles will reveal the different degrees of freedom hidden in a potentially clumsy choice of variables. These degrees of freedom can be made explicit in the action, in some cases, after the kinetic terms have been canonically normalized (e.g. via a conformal transformation).

Having identified the degrees of freedom of the theories, one needs to classify their dynamics. For this purpose, there exist two distinct approaches: one can either apply suitable transformations on the fields on the level of the Lagrangian in order to try to map one theory to the other or one can proceed by calculating straightforwardly the equations of motion of each of them and then checking whether they match in the end. It has to be stressed that both approaches lead to the same outcome and therefore we can conveniently choose how to proceed accordingly to the theory at hand.

In our previous example, we have shown that Eq. (2.2) implies that $f(R)$ theories can be described by a scalar field minimally coupled to general relativity. This means that $f(R)$ is formally equivalent to general relativity in the presence of a scalar field. In fact, both theories have the same degrees of freedom and their actions can be mapped into each other via redefinitions of the metric. As can be seen from (2.2), it is just a matter of choice whether the new scalar field ϕ belongs to the gravity sector or to the matter one.

The same logic can be used for more general theories where it is also possible to determine new degrees of freedom in addition to the metric and the scalar appearing in Eq. (2.2). In fact, an extra massive spin-2 is present in the generic theory $f(R, R_{\mu\nu}R^{\mu\nu}, R_{\mu\nu\rho\sigma}R^{\mu\nu\rho\sigma})$ [8–10]. As this is a crucial example for our investigation, we now reproduce this notorious fact using the results of Hindawi et al. [11]. Consider the theory

$$\begin{aligned} S &= \frac{1}{2\kappa^2}\int \mathrm{d}^4x\sqrt{-g}\Big[R + \alpha R^2 + \beta R_{\mu\nu}R^{\mu\nu} + \gamma R_{\lambda\mu\nu\rho}R^{\lambda\mu\nu\rho} \\ &\qquad + \mathcal{L}_M(g_{\mu\nu}, \phi_\alpha)\Big], \\ &= \frac{1}{2\kappa^2}\int d^4x\sqrt{-g}\left[R + \frac{1}{6{m_0}^2}R^2 - \frac{1}{2{m_2}^2}C^2 + \mathcal{L}_M(g_{\mu\nu}, \phi_\alpha)\right], \end{aligned} \tag{2.7}$$

where $C_{\mu\nu\rho\sigma}$ is the Weyl tensor, $m_0^{-2} = 6\alpha + 2\beta + 2\gamma$ and $m_2^{-2} = -\beta - 4\gamma$. The matter sector is represented by $\mathcal{L}_M(g_{\mu\nu}, \phi_\alpha)$, where ϕ_α denotes a set of arbitrary fields of arbitrary spin, but for the sake of the argument we will ignore the matter Lagrangian for a while. We now introduce an auxiliary scalar field λ:

$$\begin{aligned} S &= \frac{1}{2\kappa^2}\int d^4x\sqrt{-g}\left[R + \frac{1}{6{m_0}^2}R^2 - \frac{1}{6{m_0}^2}(R - 3m_0^2\lambda)^2 - \frac{1}{2{m_2}^2}C^2\right] \quad (2.8) \\ &= \frac{1}{2\kappa^2}\int d^4x\sqrt{-g}\left[(1+\lambda)R - \frac{3}{2}m_0^2\lambda^2 - \frac{1}{2{m_2}^2}C^2\right] \\ &= \frac{1}{2\kappa^2}\int d^4x\sqrt{-g}\left[e^\chi R - \frac{3}{2}m_0^2(e^\chi - 1)^2 - \frac{1}{2{m_2}^2}C^2\right]. \end{aligned}$$

In the last line, we made the redefinition $\chi = \log(1+\lambda)$. The equation of motion for λ is algebraic and given by $R = 3m_0^2\lambda$. Substituting this back into the action gives the original theory back. Hence, both models are equivalent. Now we can perform a conformal transformation $\tilde{g}_{\mu\nu} = e^\chi g_{\mu\nu}$

$$S = \frac{1}{2\kappa^2}\int d^4x\sqrt{-\tilde{g}}\left[\tilde{R} - \tfrac{3}{2}\left(\tilde{\nabla}\chi\right)^2 - \tfrac{3}{2}{m_0}^2\left(1 - e^{-\chi}\right)^2 - \frac{1}{2{m_2}^2}\tilde{C}^2\right], \tag{2.9}$$

where we have used the fact that C^2 is invariant under conformal transformations. Now we can rewrite the above action as

$$S = \frac{1}{2\kappa^2}\int d^4x\sqrt{-\tilde{g}}\Big[\tilde{R} - \tfrac{3}{2}\left(\tilde{\nabla}\chi\right)^2 - \tfrac{3}{2}m_0{}^2\left(1-e^{-\chi}\right)^2 - \frac{1}{2m_2^2}\left(\tilde{R}_{\lambda\mu\nu\rho}\tilde{R}^{\lambda\mu\nu\rho} - 2\tilde{R}_{\mu\nu}\tilde{R}^{\mu\nu} + \tfrac{1}{3}\tilde{R}^2\right)\Big] \tag{2.10}$$

$$= \frac{1}{2\kappa^2}\int d^4x\sqrt{-\tilde{g}}\Big[\tilde{R} - \tfrac{3}{2}\left(\tilde{\nabla}\chi\right)^2 - \tfrac{3}{2}m_0{}^2\left(1-e^{-\chi}\right)^2 - \frac{1}{m_2{}^2}\left(\tilde{R}_{\mu\nu}\tilde{R}^{\mu\nu} - \tfrac{1}{3}\tilde{R}^2\right) - \frac{1}{2m_2^2}\left(\tilde{R}_{\lambda\mu\nu\rho}\tilde{R}^{\lambda\mu\nu\rho} - 4\tilde{R}_{\mu\nu}\tilde{R}^{\mu\nu} + \tilde{R}^2\right)\Big]. \tag{2.11}$$

Due to the Gauss-Bonnet theorem, the last term in the last line vanishes and we end up with

$$S = \frac{1}{2\kappa^2}\int d^4x\sqrt{-\tilde{g}}\Big[\tilde{R} - \tfrac{3}{2}\left(\tilde{\nabla}\chi\right)^2 - \tfrac{3}{2}m_0{}^2\left(1-e^{-\chi}\right)^2 - \frac{1}{m_2{}^2}\left(\tilde{R}_{\mu\nu}\tilde{R}^{\mu\nu} - \tfrac{1}{3}\tilde{R}^2\right)\Big]. \tag{2.12}$$

We then include an auxiliary symmetric tensor field $\tilde{\pi}_{\mu\nu}$:

$$S = \frac{1}{2\kappa^2}\int d^4x\sqrt{-\tilde{g}}\Big[\tilde{R} - \tfrac{3}{2}\left(\tilde{\nabla}\chi\right)^2 - \tfrac{3}{2}m_0{}^2\left(1-e^{-\chi}\right)^2 - \tilde{G}_{\mu\nu}\tilde{\pi}^{\mu\nu} + \tfrac{1}{4}m_2{}^2\left(\tilde{\pi}_{\mu\nu}\tilde{\pi}^{\mu\nu} - \tilde{\pi}^2\right)\Big]. \tag{2.13}$$

where $\tilde{\pi} = \tilde{\pi}_{\mu\nu}\tilde{G}^{\mu\nu}$ and $\tilde{G}_{\mu\nu}$ is the Einstein tensor in the Einstein frame. The $\tilde{\pi}_{\mu\nu}$ equation of motion is

$$\tilde{G}_{\mu\nu} = \tfrac{1}{2}m_2^2\left(\tilde{\pi}_{\mu\nu} - \tilde{g}_{\mu\nu}\tilde{\pi}\right), \tag{2.14}$$

which can be written in the form

$$\tilde{\pi}_{\mu\nu} = 2m_2{}^{-2}\left(\tilde{R}_{\mu\nu} - \frac{1}{6}\tilde{g}_{\mu\nu}\tilde{R}\right). \tag{2.15}$$

Substituting this equation of motion back into the action (2.13) leads to the action (2.12), thus they describe the same physics. Therefore, we have proven the equivalence between the actions (2.7) and (2.13). From the action (2.13), we can see that our original theory is equivalent to general relativity in the presence of a canonical scalar field and a non-canonical symmetric rank-2 tensor field. It is tempting to say that $\tilde{\pi}_{\mu\nu}$ is a spin-2 field, but this is not obvious at this stage. So far, $\tilde{\pi}_{\mu\nu}$ contains 10 degrees of freedom, while a massive spin-2 contains only 5. In the simplest case of a free spin-2 field $\phi_{\mu\nu}$ in a flat spacetime, such field is described by the Pauli-Fierz Lagrangian. The divergence and the trace of its equation of motion imply the conditions:

$$\partial^\mu \phi_{\mu\nu} = 0, \quad \phi = 0, \tag{2.16}$$

which constrains the number of degrees of freedom to 5. However, for a generic spin-2 field the above constraints are no longer valid, but we can still find generalized constraints so to reduce the number of degrees of freedom to 5. From the trace equation of $\tilde{g}_{\mu\nu}$ and from the divergence of the $\tilde{\pi}_{\mu\nu}$ equation of motion we find:

$$\tilde{\nabla}^\mu \left(\tilde{\pi}_{\mu\nu} - \tilde{g}_{\mu\nu}\tilde{\pi}\right) = 0, \tag{2.17}$$

$$\tilde{\pi} - m_2{}^{-2}\left[\left(\tilde{\nabla}\chi\right)^2 + 2m_0{}^2\left(1 - e^{-\chi}\right)^2\right] = 0. \tag{2.18}$$

The above conditions provide 5 constraints, thus reducing the number of degrees of freedom contained in $\tilde{\pi}_{\mu\nu}$ to 5. We can conclude that $\tilde{\pi}_{\mu\nu}$ is a pure spin-2 field. Moreover, when the theory is linearized, the above conditions reproduces Pauli-Fierz constraints and, therefore, $\tilde{\pi}_{\mu\nu}$ would give rise to a canonical spin-2 field. Thus, we managed to find a spin-2 field, even though it does not appear canonically in the Lagrangian.

To canonically normalize the field $\tilde{\pi}_{\mu\nu}$, we must perform another transformation of the metric. We start by writing the Lagrangian (2.13) in the form

$$S = \frac{1}{2\kappa^2}\int d^4x\sqrt{-\tilde{g}}\left\{\left[\left(1 + \tfrac{1}{2}\tilde{\pi}\right)\tilde{g}^{\mu\nu} - \tilde{\pi}^{\mu\nu}\right]\tilde{R}_{\mu\nu} + \tfrac{1}{4}m_2{}^2\left(\tilde{\pi}_{\mu\nu}\tilde{\pi}^{\mu\nu} - \tilde{\pi}^2\right)\right.$$
$$\left. -\tfrac{3}{2}\left(\tilde{\nabla}\chi\right)^2 - \tfrac{3}{2}m_0{}^2\left(1 - e^{-\chi}\right)^2\right]\Bigg\}. \tag{2.19}$$

To obtain a canonical Einstein-Hilbert term, we have to redefine the metric as

$$\sqrt{-\bar{g}}\bar{g}^{\mu\nu} = \sqrt{-\tilde{g}}\left[\left(1 + \tfrac{1}{2}\tilde{\pi}\right)\tilde{g}^{\mu\nu} - \tilde{\pi}^{\mu\nu}\right], \tag{2.20}$$

which leads to the transformations

$$\bar{g}^{\mu\nu} = (\det A)^{-1/2}\tilde{g}^{\mu\lambda}A^\nu_\lambda \tag{2.21}$$

$$A^\nu_\lambda = \left(1 + \tfrac{1}{2}\phi\right)\delta^\nu_\lambda - \phi^\nu_\lambda. \tag{2.22}$$

We have introduced the new notation $\phi^\nu_\mu = \tilde{\pi}^\nu_\mu$ to emphasize that the indices of $\phi_{\mu\nu}$ are raised and lowered using $\bar{g}_{\mu\nu}$, while the indices of $\tilde{\pi}_{\mu\nu}$ are raised and lowered using $\tilde{g}_{\mu\nu}$. Thus, in the new set of variables the Lagrangian reads

$$S = \frac{1}{2\kappa^2}\int d^4x\sqrt{-\bar{g}}\left[\bar{R} - \tfrac{3}{2}\left(A^{-1}(\phi_{\sigma\tau})\right)_\mu{}^\nu\bar{\nabla}^\mu\chi\bar{\nabla}_\nu\chi - \tfrac{3}{2}m_0{}^2\left(\det A(\phi_{\sigma\tau})\right)^{-1/2}\left(1 - e^{-\chi}\right)^2\right.$$
$$- \bar{g}^{\mu\nu}\left(C^\lambda{}_{\mu\rho}(\phi_{\sigma\tau})C^\rho{}_{\nu\lambda}(\phi_{\sigma\tau}) - C^\lambda{}_{\mu\nu}(\phi_{\sigma\tau})C^\rho{}_{\rho\lambda}(\phi_{\sigma\tau})\right) \tag{2.23}$$
$$\left. + \tfrac{1}{4}m_2{}^2\left(\det A(\phi_{\sigma\tau})\right)^{-1/2}\left(\phi_{\mu\nu}\phi^{\mu\nu} - \phi^2\right)\right],$$

where

$$C^{\lambda}{}_{\mu\nu} = \tfrac{1}{2}(\tilde{g}^{-1})^{\lambda\rho}(\bar{\nabla}_\mu \tilde{g}_{\nu\rho} + \bar{\nabla}_\nu \tilde{g}_{\mu\rho} - \bar{\nabla}_\rho \tilde{g}_{\mu\nu}). \tag{2.24}$$

Due to the transformation 2.21, the metric $\tilde{g} = \tilde{g}(\phi_{\mu\nu})$ now depends on the spin-2 field. Thus the spin-2 kinetic term appears explicitly in the action through $C^{\lambda}{}_{\mu\nu}$.

When external matter is present, the argument goes in the same way, except that after performing the transformations the matter Lagrangian becomes $\mathcal{L}_M(e^{-\chi}\tilde{g}_{\mu\nu}(\phi_{\sigma\tau}), \phi_\alpha)$ and the action reads

$$\begin{aligned} S = \frac{1}{2\kappa^2}\int d^4x\sqrt{-\bar{g}}\Big[&\bar{R} - \tfrac{3}{2}\left(A^{-1}(\phi_{\sigma\tau})\right)_\mu{}^\nu \bar{\nabla}^\mu\chi\bar{\nabla}_\nu\chi - \tfrac{3}{2}m_0{}^2\,(\det A(\phi_{\sigma\tau}))^{-1/2}\left(1-e^{-\chi}\right)^2 \\ &- \bar{g}^{\mu\nu}\left(C^{\lambda}{}_{\mu\rho}(\phi_{\sigma\tau})C^{\rho}{}_{\nu\lambda}(\phi_{\sigma\tau}) - C^{\lambda}{}_{\mu\nu}(\phi_{\sigma\tau})C^{\rho}{}_{\rho\lambda}(\phi_{\sigma\tau})\right) \\ &+ \tfrac{1}{4}m_2{}^2\,(\det A(\phi_{\sigma\tau}))^{-1/2}\left(\phi_{\mu\nu}\phi^{\mu\nu} - \phi^2\right) + \bar{\mathcal{L}}_M(e^{-\chi}\tilde{g}_{\mu\nu}(\phi_{\sigma\tau}), \phi_\alpha)\Big], \end{aligned} \tag{2.25}$$

where

$$\bar{\mathcal{L}}_M = e^{-2\chi}(\det A(\phi_{\mu\nu}))^{-1/2}\mathcal{L}_M. \tag{2.26}$$

We see that, in general, external matter couples minimally to the usual graviton through the Jacobian $\sqrt{-\bar{g}}$ and non-minimally to the fields χ and $\phi_{\mu\nu}$.

In the following, we calculate explicitly the coupling between external matter and the extra degrees of freedom χ and $\phi_{\mu\nu}$. Consider a matter Lagrangian composed of a scalar, a vector and a spinor field:

$$\mathcal{L}_M = \mathcal{L}_0 + \mathcal{L}_1 + \mathcal{L}_{1/2}, \tag{2.27}$$

where

$$\mathcal{L}_0 = \tfrac{1}{2}\nabla_\mu\sigma\nabla^\mu\sigma \tag{2.28}$$

$$\mathcal{L}_1 = -\tfrac{1}{4}F_{\mu\nu}F^{\mu\nu} \tag{2.29}$$

$$\mathcal{L}_{1/2} = i\bar{\psi}\partial\!\!\!/\psi. \tag{2.30}$$

After transforming the metric to $\bar{g}_{\mu\nu}$ (i.e., $g_{\mu\nu} \to \tilde{g}_{\mu\nu} \to \bar{g}_{\mu\nu}$), we obtain

$$\bar{\mathcal{L}}_0 = \tfrac{1}{2}e^{-\chi}(A^{-1})_\alpha{}^\nu\bar{g}^{\alpha\mu}\nabla_\mu\sigma\nabla_\nu\sigma, \tag{2.31}$$

$$\bar{\mathcal{L}}_1 = -\tfrac{1}{4}(\det A)^{1/2}(A^{-1})_\rho{}^\mu(A^{-1})_\lambda{}^\nu\bar{g}^{\rho\alpha}\bar{g}^{\lambda\beta}F_{\mu\nu}F_{\alpha\beta}, \tag{2.32}$$

$$\bar{\mathcal{L}}_{1/2} = e^{-\chi}(A^{-1})_\alpha{}^\nu i\bar{\psi}\bar{g}^{\alpha\mu}\gamma_\mu\partial_\nu\psi, \tag{2.33}$$

and $\bar{\mathcal{L}}_M = \bar{\mathcal{L}}_0 + \bar{\mathcal{L}}_1 + \bar{\mathcal{L}}_{1/2}$. One can also include interaction terms, namely the Yukawa interaction and the gauge interactions for spinor-vector fields and scalar-vector fields and look at how they are affected by the metric redefinition:

$$\mathcal{L}_{\text{Yukawa}} = -g\bar{\psi}\phi\psi, \tag{2.34}$$

$$\mathcal{L}_0 = \tfrac{1}{2}(D_\mu\sigma)^\dagger(D^\mu\sigma) = \frac{1}{2}\nabla_\mu\sigma\nabla^\mu\sigma + e^2A_\mu A^\mu\sigma^2, \tag{2.35}$$

$$\mathcal{L}_{1/2} = i\bar{\psi}\not{D}\psi = i\bar{\psi}\gamma^\mu\nabla_\mu\psi - eA_\mu\bar{\psi}\gamma^\mu\psi, \tag{2.36}$$

where $D_\mu = \nabla_\mu + ieA_\mu$. After transforming the metric to $\bar{g}_{\mu\nu}$ (i.e., $g_{\mu\nu} \to \tilde{g}_{\mu\nu} \to \bar{g}_{\mu\nu}$), one finds

$$\bar{\mathcal{L}}_{\text{Yukawa}} = -e^{-2\chi}(\det A)^{-1/2}g\bar{\psi}\phi\psi, \tag{2.37}$$

$$\bar{\mathcal{L}}_0 = \tfrac{1}{2}e^{-\chi}(A^{-1})_\alpha{}^\nu\bar{g}^{\alpha\mu}(\nabla_\mu\sigma\nabla_\nu\sigma + e^2A_\mu A_\nu\sigma^2), \tag{2.38}$$

$$\bar{\mathcal{L}}_{1/2} = e^{-\chi}(A^{-1})_\alpha{}^\nu\bar{g}^{\alpha\mu}(i\bar{\psi}\gamma_\mu\partial_\nu\psi - eA_\mu\bar{\psi}\gamma_\nu\psi), \tag{2.39}$$

and $\bar{\mathcal{L}}_M = \bar{\mathcal{L}}_0 + \bar{\mathcal{L}}_1 + \bar{\mathcal{L}}_{1/2} + \bar{\mathcal{L}}_{\text{Yukawa}}$. We note that the massive spin-2 field couples to all matter fields of spin 0, 1/2 and 2 because of the matrix A. On the other hand, the scalar field χ does not couple to photons. The masses of the spin-0 and massive spin-2 fields can be tuned by adjusting the coefficients of the action. However, their interactions with matter fields, while not always universal, are determined by the gravitational coupling constant. As usual, the massless graviton couples universally and gravitationally to matter fields.

2.3 Application to Dark Matter

As already stressed, astrophysical and cosmological evidences for dark matter are overwhelming. Many models have been proposed to explain the dark matter phenomenon. These models are usually classified into two classes: modifications of Einstein's general relativity or modifications of the standard model in the form of new particles. The purpose of this section is to point out that these two categories are not so different after all. In fact, every modified gravity model has new degrees of freedom in addition to the usual massless graviton.

The first attempt to explain galaxy rotation curves by a modification of Newtonian dynamics is due to Milgrom [12]. While Milgrom's original proposal was non-relativistic and very phenomenological, more refined theories have been proposed later on, including Bekenstein's TeVeS theory [13], Moffat's modified gravity (MOG) [14] and Mannheim's conformal gravity [15], which are relativistic. While these theories seem to be able to explain the rotation curves of galaxies (see e.g. [16] for a recent MOND review where the observational successes are discussed in details), it is more challenging to imagine how they would explain the bullet cluster observations or the agreement of the CMB observation with the standard cosmological model ΛCMB which posits the existence of cold dark matter. We shall not dwell on the question of the viability of modified gravity as we may simply not yet have found the right model. Nonetheless, we merely point out that if such a theory exists, it will not be necessarily distinct from a theory involving particles as dark matter.

In fact, whatever this realistic theory turns out to be, it can always be parameterized by a function $f(R, R_{\mu\nu}, R_{\mu\nu\rho\sigma}, \phi_\alpha)$ modelled using effective theory techniques. Here R is the Ricci scalar, $R_{\mu\nu}$ is the Ricci tensor and ϕ_α denotes collectively any type of field that is also responsible for the gravitational interaction. In terms of effective field theory, any theory of modified gravity can be described by

$$S = \frac{1}{16\pi G} \int d^4x \sqrt{-g} f(R, R_{\mu\nu}, R_{\mu\nu\rho\sigma}, \phi_\alpha) + \int d^4x \sqrt{-g} \mathcal{L}_M \quad (2.40)$$

where G is Newton's constant. We are only assuming diffeomorphism invariance and the usual space-time and gauge symmetries for the matter content described by the Lagrangian $\mathcal{L}_M$. A successful model should lead to a modification of Newton's potential that fits, e.g., the galaxy rotation curves. It is not hard to imagine that the standard Newtonian term $1/r$ would come from the usual massless spin-2 graviton exchange while the non-Newtonian terms would have to be generated by the new degrees of freedom. Clearly, it is not straightforward to come up with such a model, however, as mentioned before, there are a few known examples.

While it is clear that new degrees of freedom are included when ϕ_α is included to the function f as in Moffat [14], it is much less obvious how they are determined when the theory is a function of the curvature invariants only as we emphasized before. Thus, we will restrict ourselves to the theory $f(R, R_{\mu\nu}, R_{\mu\nu\rho\sigma})$. From the arguments made at the end of Sect. 2.2, we know that this model is equivalent to general relativity in the presence of massive spin-0 and spin-2 fields. We conclude that there is no difference between introducing new particles and introducing modifications of gravity, which raises the question of whether it is possible to distinguish experimentally between models of modified gravity and particle dark matter.

Any modification of gravity that has the diffeomorphism group as the symmetry group can be reformulated, using appropriate variables, as usual general relativity accompanied by new fields. We have seen that these new degrees of freedom may not couple universally to matter. Therefore, modified gravity can be seen as a model with new dark matter particles that are very weakly coupled to the standard model sector. These apparently very different models describe the same physics as their actions are related by simple field redefinitions. This may provide a simple way for modified gravity proponents to support bullet cluster experiments or the cosmic microwave background.

2.4 Conclusions

In this chapter, we proposed a classification strategy for gravitational theories. In particular, we have demonstrated the equivalence of the broad class of theories $f(R, R_{\mu\nu}, R_{\mu\nu\rho\sigma})$ and general relativity in the company of extra matter fields, namely a massive spin-0 and a massive spin-2 fields. We have shown that these new degrees of freedom can be shifted into a redefined energy-momentum tensor and that they couple universally to the matter fields introduced in the model. We conclude that any attempt to modify the Einstein-Hilbert action, preserving the underlying symmetry, leads to new degrees of freedom, i.e., new particles. In that sense, this is not different from adding new matter fields by hand to the matter sector that are coupled gravitationally to the standard model. Assuming that theories of modified gravity preserve diffeomorphism invariance, we have argued that they are equivalent to general relativity with new degrees of freedom coupled gravitationally to the fields of the standard model. From this point of view, there is a duality between models of modified gravity and particle physics models with new fields that are coupled gravitationally to the standard model.

These results may ease the analysis of the physics of dark matter models involving modifications of gravity and, in particular, the fact that they are dual to some very weakly coupled dark matter model could help to resolve the apparent disagreement with the bullet cluster observation.

We note that, while we focussed on dark matter as an application of the gravitational theories classification that we have proposed, there should be important consequences to other phenomena such as dark energy, inflation, gravitational waves etc, that we have not discussed.

References

1. Hooper D (2010) Particle dark matter. In Proceedings of Theoretical Advanced Study Institute in Elementary Particle Physics on The dawn of the LHC era (TASI 2008): Boulder, USA, June 2–27, 2008, pp 709–764
2. Chen P, Ong YC, Yeom D-H (2015) Black hole remnants and the information loss paradox. Phys Rept 603:1–45
3. Calmet X (2015) Virtual black holes, remnants and the information paradox. Class Quant Grav 32(4):045007
4. Clifton T, Ferreira PG, Padilla A, Skordis C (2012) Modified gravity and cosmology. Phys Rept 513:1–189
5. Calmet X, Yang T-C (2013) Frame transformations of gravitational theories. Int J Mod Phys A 28:1350042
6. Magnano G (1995) Are there metric theories of gravity other than general relativity? Gen Relativ Gravitational Phys. In: Proceedings, 11th Italian Conference. Trieste, Italy, September 26–30, 1994, pp 213–234
7. De Felice A, Tsujikawa S (2010) f(R) theories. Living Rev Rel 13:3
8. Magnano G, Sokolowski LM (2003) Nonlinear massive spin two field generated by higher derivative gravity. Annals Phys 306:1–35
9. Nunez A, Solganik S (2005) Ghost constraints on modified gravity. Phys Lett B 608:189–193
10. Chiba T (2005) Generalized gravity and ghost. JCAP 0503:008
11. Hindawi A, Ovrut BA, Waldram D (1996) Consistent spin two coupling and quadratic gravitation. Phys Rev D 53:5583–5596
12. Milgrom M (1983a) A modification of the Newtonian dynamics as a possible alternative to the hidden mass hypothesis. Astrophys J 270:365–370

13. Bekenstein JD (2004) Relativistic gravitation theory for the MOND paradigm. Phys Rev D 70:083509. (Erratum: Phys Rev D 71:069901, 2005)
14. Moffat JW (2006) Scalar-tensor-vector gravity theory. JCAP 0603:004
15. Mannheim PD (2012) Making the case for conformal gravity. Found Phys 42:388–420
16. Famaey B, McGaugh S (2012) Modified Newtonian dynamics (MOND): observational phenomenology and relativistic extensions. Living Rev Rel 15:10

Chapter 3
Gravitational Waves in Quantum Gravity

In this chapter, we look at quantum effects on Einstein's equations using the effective field theory formalism. We take into account the leading order quantum gravitational correction to the wave equation. In addition to the massless mode corresponding to the graviton, we find another two modes with complex masses. The imaginary part of these masses are interpreted as a width and, therefore, could produce a damping of gravitational waves should these modes be excited in strong astrophysical processes, such as black hole mergers, creating gravitational waves. We analyze the significance for gravitational wave events such as GW 150914 recently observed by the Advanced LIGO collaboration.

3.1 Introduction

The recent discovery of gravitational waves by the Advanced LIGO collaboration [1] marks the beginning of a new era in astronomy which could shed some new light on our universe, uncovering its dark elements that do not interact with electromagnetic radiations. This finding could also yield some new insights in theoretical physics. In this chapter, we investigate the leading order quantum gravitational effects of gravitational waves using the effective field theory framework. Although a complete theory of quantum gravity remains unknown, it is possible to make use of effective field theory techniques to make actual predictions of quantum gravity. Assuming that diffeomorphism invariance is the right symmetry of the gravitational interaction at the Planck scale and supposing that we have knowledge of the field content below the Planck scale, one can write down an effective action for quantum gravity that is independent of our knowledge of physics in the UV. This effective theory is valid up to energies close to the Planck scale. It can be obtained by linearizing general relativity around a chosen background. The graviton is described by a massless spin-2 tensor which is quantized using the usual quantum field theoretical methods. This theory

© Springer Nature Switzerland AG 2019

I. Kuntz, *Gravitational Theories Beyond General Relativity*,
Springer Theses, https://doi.org/10.1007/978-3-030-21197-4_3

is known to be non-renormalizable, but the divergences can be absorbed into the Wilson coefficients of higher dimensional operators compatible with diffeomorphism symmetry. The difference with respect to a usual renormalizable theory lies in the fact that an infinite number of measurements are necessary to determine the action to all orders. However, only a finite number of coefficients are undetermined below the Planck scale, allowing for model-independent predictions of quantum gravity which serve as true tests of the quantum nature of gravity, regardless of physics in the UV.

We will first study quantum gravitational corrections to the linearized Einstein's equations, which will ultimately lead to corrections to the wave equation. By solving these equations, we show that in addition to the standard solution that corresponds to the massless graviton, there are solutions corresponding to massive particles. Should these massive particles be excited during astrophysical events, a considerable fraction of the energy released by such events could be emitted into these modes. We will show that the gravitational waves corresponding to these modes are damped, thus the energy of the wave could be dissipated. We then investigate whether the recent discovery of gravitational waves by the Advanced LIGO collaboration [1] could provide a test of quantum gravity.

3.2 Resummed Graviton Propagator and Gravitational Waves

Given a matter Lagrangian coupled to general relativity with N_s scalar degrees of freedom, N_f fermions and N_V vectors, the graviton vacuum polarization can be calculated in the large $N = N_s + 3N_f + 12N_V$ limit while keeping NG_N small, where G_N is Newton's constant. Since we are interested in energies below $M_\star$, where the effective field theory supposedly breaks down, we need not take into account the graviton self-interactions, which are suppressed by powers of $1/N$ in comparison to the matter loops. Observe that $M_\star$ is a dynamical object and does not necessarily agree with the reduced Planck mass of order 10^{18} GeV (see e.g. [2]). The divergence in this diagram can be isolated using dimensional regularization and absorbed in the coefficient of R^2 and $R_{\mu\nu}R^{\mu\nu}$. An infinite series of vacuum polarization diagrams contributing to the dressed graviton propagator can be resummed in the large N limit. This technique leads to the resummed graviton propagator given by Aydemir et al. [3]

$$iD^{\alpha\beta,\mu\nu}(q^2) = \frac{i\left(L^{\alpha\mu}L^{\beta\nu} + L^{\alpha\nu}L^{\beta\mu} - L^{\alpha\beta}L^{\mu\nu}\right)}{2q^2\left(1 - \frac{NG_Nq^2}{120\pi}\log\left(-\frac{q^2}{\mu^2}\right)\right)} \tag{3.1}$$

where $L^{\mu\nu}(q) = \eta^{\mu\nu} - q^\mu q^\nu/q^2$, q^μ is the 4-momentum, μ is the renormalization scale and the $i\varepsilon$ prescription is implicit. This resummed propagator is the source of interesting acausal and non-local phenomena which have just started to be investigated [3–8]. Here we will focus on how gravitational waves are affected by these quantum gravity effects.

From the dressed graviton propagator in momentum space, we can easily read off the classical field equation for the spin-2 gravitational wave in momentum space

$$2q^2\left(1-\frac{NG_Nq^2}{120\pi}\log\left(-\frac{q^2}{\mu^2}\right)\right)=0. \tag{3.2}$$

This equation has three solutions [7]:

$$\begin{aligned} q_1^2 &= 0, \\ q_2^2 &= \frac{1}{G_N N}\frac{120\pi}{W\left(\frac{-120\pi}{\mu^2 N G_N}\right)}, \\ q_3^2 &= (q_2^2)^*, \end{aligned} \tag{3.3}$$

where W is the Lambert function. The complex pole corresponds to a new massive degree of freedom with a complex mass (i.e. they have a width [7]). The general wave solution is thus of the form

$$h^{\mu\nu}(x) = a_1^{\mu\nu}\exp(-iq_{1\alpha}x^\alpha) + a_2^{\mu\nu}\exp(-iq_{2\alpha}x^\alpha) + a_3^{\mu\nu}\exp(-iq_{2\alpha}^\star x^\alpha), \tag{3.4}$$

where $a_i^{\mu\nu}$ are polarization tensors. Hence, there are three degrees of freedom that can be excited in gravitational processes, leading to gravitational wave emissions. Note that our solution is linear, non-linearities in gravitational waves (see e.g. [9]) have been studied and are, as expected, negligible.

The position of the complex pole depends on the number of fields present in the model. In the standard model of particle physics, one has $N_s = 4$, $N_f = 45$, and $N_V = 12$. We thus find $N = 283$ and the pair of complex poles at $(7 - 3i) \times 10^{18}$ GeV and $(7 + 3i) \times 10^{18}$ GeV. Note that the pole q_3^2 corresponds to a particle with an incorrect sign between the squared mass and the width term, i.e. a ghost. We will not study this Lee-Wick pole any further and will assume that this issue is healed by strong gravitational interactions at the UV. The renormalization scale needs to be adjusted to match the number of particles included in the model. Indeed, to a good approximation the real part of the complex pole is of the order of

$$|\text{Re}\, q_2| \sim \sqrt{\frac{120\pi}{NG_N}} \tag{3.5}$$

which corresponds to the energy scale $M_\star$ at which the effective theory breaks down. In fact, the complex pole will produce acausal effects, signaling strong quantum gravitational effects which cannot be described within the domain of the effective theory. Therefore, we should choose the renormalization scale μ of the order of $M_\star \sim |\text{Re}\, q_2|$. We have

$$q_2^2 \approx \pm \frac{1}{G_N N} \frac{120\pi}{W(-1)} \approx \mp (0.17 + 0.71\, i) \frac{120\pi}{G_N N}, \tag{3.6}$$

and we thus find the mass of the complex pole:

$$m_2 = (0.53 - 0.67\, i) \sqrt{\frac{120\pi}{G_N N}}. \tag{3.7}$$

As stressed before, the mass of this object depends on the number of fields in the theory. The corresponding wave has a frequency:

$$\begin{aligned} w_2 = q_2^0 &= \pm \sqrt{\vec{q}_2 . \vec{q}_2 + (0.17 + 0.71\, i) \frac{120\pi}{G_N N}} \qquad (3.8)\\ &= \pm \left(\frac{1}{\sqrt{2}} \sqrt{\sqrt{\left(\vec{q}_2 . \vec{q}_2 + 0.17 \frac{120\pi}{G_N N} \right)^2 + \left(0.71 \frac{120\pi}{G_N N} \right)^2} + \vec{q}_2 . \vec{q}_2 + 0.17 \frac{120\pi}{G_N N}} \right. \\ &\left. + i \frac{1}{\sqrt{2}} \sqrt{\sqrt{\left(\vec{q}_2 . \vec{q}_2 + 0.17 \frac{120\pi}{G_N N} \right)^2 + \left(0.71 \frac{120\pi}{G_N N} \right)^2} - \vec{q}_2 . \vec{q}_2 - 0.17 \frac{120\pi}{G_N N}} \right). \end{aligned}$$

The imaginary part of the complex pole will produce a damping of the component of the gravitational wave associated to that mode. Since the complex poles are also coupled to matter, we must assume that the massive modes are produced at the same rate as the massless graviton mode, should this be allowed kinematically. During an astrophysical process leading to gravitational waves, part of the energy will be emitted into these massive modes, causing them to decay rather rapidly because of their huge decay width. The possible damping of the gravitational wave suggests that we should be careful when relating the gravitational wave energy measured on earth to that of an astrophysical event since part of this energy could have been dissipated away as the wave travels towards the earth.

The idea that gravitational waves could experience some damping has been previously considered [10]. Nonethesless, it is well known that the graviton cannot be split into many gravitons, even at the quantum level [11]. Should there be such an effect, it would have to be at the non-perturbative level [12]. In our case, the massless mode is not damped, thus there is no contradiction with the work of Fiore and Modanese [11]. Moreover, as already stressed, the dispersion relation of the massless mode of the gravitational wave is not affected, we do not violate any fundamental symmetry such as Lorentz invariance. This is in contrast to the model presented in Arzano and Calcagni [13].

Since the complex poles couple to matter with the same coupling as the standard massless graviton, they can be thought of as a massive graviton, even though strictly speaking these modes have two polarizations only as opposed to massive gravitons that have five. This idea has been applied in the context of $F(R)$ gravity [14] (see also [15, 16] for previous works on gravitational waves in $F(R)$ gravity). We will assume that these massive modes can be excited during the coalescence of two black holes. As a first approximation, we assume that all the energy released during the coalescence is emitted into these modes. Then one can use the bound derived by the LIGO collaboration on a graviton mass. We know that $m_g < 1.2 \times 10^{-22}$ eV [1] and we can thus get a constraint:

$$\sqrt{\mathrm{Re}\left(\frac{1}{G_N N}\frac{120\pi}{W\left(\frac{-120\pi M_P^2}{\mu^2 N}\right)}\right)} < 1.2 \times 10^{-22}\ \mathrm{eV}. \tag{3.9}$$

Therefore, we get a lower bound on N: $N > 4 \times 10^{102}$ considering that all the energy of the merger was carried out by the massive degrees of freedom. Evidently, this situation is not very realistic as the massless degree of freedom will also be excited. Nonetheless, it suggests that, should the massive modes be produced, they would only reach the earth for masses smaller than 1.2×10^{-22} eV. Gravitational waves associated to heavier poles will be damped before arriving on earth. We will see below that there are more constraining bounds on the mass of these modes coming from Eötvös type pendulum experiments.

We now need to discuss which degrees of freedom could have been produced during the coalescence of the two black holes that led to the gravitational waves measured by the LIGO collaboration. The LIGO collaboration determines that the gravitational wave event GW150914 was produced by the merger of two black holes: the black holes follow an inspiral orbit before merging and subsequently going through a final black hole ringdown. Over 0.2 s, the signal increases in frequency and amplitude in about 8 cycles from 35 to 150 Hz, where the amplitude reaches a maximum [1]. The typical energy of the gravitational wave is of the order of 150 Hz or 6×10^{-13} eV. In other words, if the gravitational wave had been emitted in the massive mode, they could not have been more massive than 6×10^{-22} GeV. Nonetheless, this indicates that it is perfectly possible that a considerable number of massive modes with $m_g < 1.2 \times 10^{-22}$ eV could have been produced.

Let us now revisit the constraint on the number of fields N and the new complex pole using Eötvös type pendulum experiments by looking for departures from the Newtonian $1/r$ potential. The dressed graviton propagator discussed above can be represented by the effective operator

$$\frac{N}{2304\pi^2} R \log\left(\frac{\Box}{\mu^2}\right) R \tag{3.10}$$

where R is the Ricci scalar. As explained above, the log term will be a contribution of order 1, this operator is thus very similar to the more familiar cR^2 term studied by Stelle. The current bound on the Wilson coefficient of c is $c < 10^{61}$ [17–19]. We can express this bound as a bound on N: $N < 2 \times 10^{65}$. This indicates that the mass of the complex pole must be larger than 5×10^{-13}GeV. This constraint, although very weak, is tighter than the one we have obtained from the graviton mass by 37 orders of magnitude.

3.3 Conclusions

In this chapter, we have studied the effect of quantum gravity on gravitational waves using effective field theory techniques, which are independent of UV completions. We discovered that quantum gravity yields new poles in the dressed propagator of the graviton in addition to the standard massless degree of freedom corresponding to the graviton. These new poles are massive and couple gravitationally to matter. Therefore, if kinematically allowed, they would be produced in approximately the same amount as the massless degree of freedom in astrophysical processes. A considerable portion of the energy released in astrophysical processes could then be carried away by massive degrees of freedom, which would decay and produce a damping of the corresponding component of the gravitational wave. While our back-of-the-envelope calculation suggests that the energy produced in the coalescence recently observed by LIGO was unlikely to be sufficiently high to generate such modes, one should be careful in extrapolating the amount of energy of astrophysical processes from the energy of the gravitational wave measured on earth. This could be a major effect for primordial gravitational waves should the scale of inflation be in the region of 10^{16} GeV, i.e. within a few orders of magnitude of the Planck scale.

References

1. Abbott BP et al (2016) Observation of gravitational waves from a binary black hole merger. Phys Rev Lett 116(6):061102
2. Calmet X (2013) Effective theory for quantum gravity. Int J Mod Phys D 22:1342014
3. Aydemir U, Anber MM, Donoghue JF (2012) Self-healing of unitarity in effective field theories and the onset of new physics. Phys Rev D 86:014025
4. Donoghue JF, El-Menoufi BK (2014) Nonlocal quantum effects in cosmology: Quantum memory, nonlocal FLRW equations, and singularity avoidance. Phys Rev D 89(10):104062
5. Calmet X, Casadio R (2014) Self-healing of unitarity in Higgs inflation. Phys Lett B 734:17–20
6. Calmet X, Croon D, Fritz C (2015) Non-locality in quantum field theory due to general relativity. Eur Phys J C 75(12):605
7. Calmet X (2014) The lightest of black holes. Mod Phys Lett A 29(38):1450204
8. Calmet X, Casadio R (2015) The horizon of the lightest black hole. Eur Phys J C 75(9):445
9. Aldrovandi R, Pereira JG, da Rocha R, Vu KH (2010) Nonlinear gravitational waves: their form and effects. Int J Theor Phys 49:549–563
10. Jones P, Singleton D (2015) Gravitons to photons—Attenuation of gravitational waves. Int J Mod Phys D 24(12):1544017
11. Fiore G, Modanese G (1996) General properties of the decay amplitudes for massless particles. Nucl Phys B 477:623–651
12. Efroimsky M (1994) Weak gravitation waves in vacuum and in media: Taking nonlinearity into account. Phys Rev D 49:6512–6520
13. Arzano M, Calcagni G (2016) What gravity waves are telling about quantum spacetime. Phys Rev D 93(12):124065. (Addendum: Phys Rev D 94(4):049907 2016)
14. Vainio J, Vilja I (2017) $f(R)$ gravity constraints from gravitational waves. Gen Rel Grav 49(8):99
15. Bogdanos C, Capozziello S, De Laurentis M, Nesseris S (2010) Massive, massless and ghost modes of gravitational waves from higher-order gravity. Astropart Phys 34:236–244

16. Capozziello S, Stabile A (2015) Gravitational waves in fourth order gravity. Astrophys Space Sci 358(2):27
17. Hoyle CD, Kapner DJ, Heckel BR, Adelberger EG, Gundlach JH, Schmidt U, Swanson HE (2004) Sub-millimeter tests of the gravitational inverse-square law. Phys Rev D 70:042004
18. Stelle KS (1978) Classical gravity with higher derivatives. Gen Rel Grav 9:353–371
19. Calmet X, Hsu SDH, Reeb D (2008) Quantum gravity at a TeV and the renormalization of Newton's constant. Phys Rev D 77:125015

Chapter 4
Backreaction of Quantum Gravitational Modes

Effective Field Theory methods are employed to investigate the leading order quantum corrections to the backreaction of gravitational waves. The effective energy-momentum tensor is computed and we show that it has a non-zero trace that contributes to the cosmological constant. By confronting our result with LIGO's data, the first constraint on the amplitude of the massive mode is obtained: $\epsilon < 1.4 \times 10^{-33}$.

4.1 Introduction

The recent experimental discovery of gravitational waves (GWs) [1] has marked a new era for both observational and theoretical physics. With the new forthcoming data from LIGO and from future experiments such as LISA, it will become possible to put modified gravity through its paces, opening up the possibility of determining for which range of parameters these theories agree with observations (if they agree at all). In particular, it might be even possible to test quantum gravity in the infrared regime, even though a UV completed theory for quantum gravity persists as one of the most important problems in modern physics.

An obvious observable to be considered is the GW energy. As a non-linear phenomenon, gravity couples to itself and thus gravitates, meaning that GWs—being a manifestation of gravity itself—induce a backreaction onto the spacetime. Therefore, we should be able to find an energy-momentum tensor for the GWs that could account for this effect. In the case of classical General Relativity (GR), such a energy-momentum tensor is known:

© Springer Nature Switzerland AG 2019

I. Kuntz, *Gravitational Theories Beyond General Relativity*,
Springer Theses, https://doi.org/10.1007/978-3-030-21197-4_4

$$t_{\mu\nu}^{\text{GR}} = \frac{1}{32\pi G} \langle \partial_\mu h_{\alpha\beta} \partial_\nu h^{\alpha\beta} \rangle, \tag{4.1}$$

where $h_{\mu\nu}$ are the perturbations of the metric and the brackets is an average over spacetime, which is responsible for taking only the long-wavelength modes into account; its rigorous definition will be introduced later on. The GW energy-momentum tensor has also been computed for some other theories, including $f(R)$, Chern-Simons and higher-derivative gravity [2–5]. In [6], it was shown how the parameters of an analytic $f(R)$ theory could be bounded by the observation of the energy or momentum carried away by the GWs.

The phenomenology, however, is not our only motivation. An alternative for the origin of dark energy has been proposed based on the effective energy-momentum tensor [4, 7–9]. Although this is not possible in GR as a result of the vanishing trace of $t_{\mu\nu}^{\text{GR}}$, it has been pointed out that it could be possible in modified gravity theories. Nonetheless, it has also been found that in some theories such as Starobinsky gravity, the effective energy-momentum tensor could not be the only factor as it does not yield the correct value for the cosmological constant [4]. We will show that the huge contributions from the Standard Model cannot be canceled by the quantum gravitational effects, thus demanding the existence of some other mechanism capable of reconciling the discrepancy between theory and observation.

The goal of this chapter is thus two-fold: we will set up new phenomenological bounds and discuss the possibility of producing a contribution to the cosmological constant in this formalism. Effective Field Theory methods will be used to compute the quantum contributions to the GW backreaction and to the wave equation in an arbitrary background. The short-wave formalism will be used, which consists of an averaging operation that splits the low-frequency modes from the high-frequency ones, so to compute the GW energy-momentum tensor in quantum GR. These theoretical findings will be helpful to constrain some of the parameters of Effective Quantum Gravity by comparing with LIGO's observations. Moreover, on the theoretical side, they give us new insights into gravity at the quantum level since this approach is independent of any UV completion and, as such, leads to genuine predictions of Quantum Gravity.

This chapter is structured as follows. In Sect. 4.2, we will review some of the most important results of the Effective Field Theory of gravity. In Sect. 4.3, we employ the short-wave formalism to compute the leading order quantum corrections to the GW energy-momentum tensor. The result will allow us to obtain a bound on the amplitude of the massive mode present in Effective Quantum Gravity. In Sect. 4.4, we discuss the quantum corrections to the propagation of GWs and we show that the equation describing the propagation in curved spacetime can be obtained by carrying out a minimal coupling recipe to the equation in Minkowski space. We draw the conclusions in Sect. 4.5.

4.2 Effective Quantum Gravity

The quantum effective action of gravity up to second order in curvature reads [10]

$$\Gamma = \int d^4x\sqrt{-g}\left(\frac{M_p^2}{2}R + b_1 R^2 + b_2 R_{\mu\nu}R^{\mu\nu} + c_1 R \log\frac{-\Box}{\mu^2}R \right.$$
$$\left. + c_2 R_{\mu\nu} \log\frac{-\Box}{\mu^2}R^{\mu\nu} + c_3 R_{\mu\nu\rho\sigma} \log\frac{-\Box}{\mu^2}R^{\mu\nu\rho\sigma}\right), \tag{4.2}$$

where $M_p = (8\pi G)^{-1/2}$ is the reduced Planck mass, G is the Newton's constant, μ is the renormalization scale and the kernel R denotes the Riemann tensor and its contractions (Ricci tensor and Ricci scalar) depending on the number of indices it carries. We will use the signature $(-+++)$. We set the bare cosmological constant to zero as it is not significant to our discussions. The coefficients b_i are free parameters and must be determined by observations, while the coefficients c_i are real predictions of the infra-red theory and only depend on the field content under consideration (see Table 1 in [10] for their precise values). The log operators lead to acausal effects that must be eliminated by representing the non-local operator as

$$\log\frac{-\Box}{\mu^2} = \int_0^\infty ds\left(\frac{1}{\mu^2 + s} - G(x, x', \sqrt{s})\right), \tag{4.3}$$

where $G(x, x'; \sqrt{s})$ is a Green's function for

$$(-\Box + k^2)G(x, x'; k) = \delta^4(x - x'), \tag{4.4}$$

and imposing retarded boundary conditions on $G(x, x'; k)$ in order to obtain a result that respects causality. Futhermore, in the limit where the wavelength is much smaller than the curvature radius $\lambda/L \ll 1$ (see Sect. 4.3), the log terms are not independent because of the following relation (see [3]):

$$\delta \int d^4x\sqrt{-g}\left(R_{\mu\nu\rho\sigma}\log\frac{-\Box}{\mu^2}R^{\mu\nu\rho\sigma} - 4R_{\mu\nu}\log\frac{-\Box}{\mu^2}R^{\mu\nu} + R\log\frac{-\Box}{\mu^2}R\right) \overset{\text{weak}}{=} 0. \tag{4.5}$$

This can also be seen by linearizing the field equations as performed in [11]. It is important to stress that the log operators in the above formula certainly violate the topological invariance of the Gauss-Bonnet theorem. However, such relation still provides a very helpful formula that can be put in use to simplify calculations in the regime of small wavelengths. Hence, since we are only interested in the scenario where $\lambda/L \ll 1$, the last term in (4.2) will be eliminated in favour of the other two log terms, translating into a redefinition of their coefficients:

$$c_1 \to \alpha \equiv c_1 - c_3, \tag{4.6}$$

$$c_2 \to \beta \equiv c_2 + 4c_3. \tag{4.7}$$

Therefore, from now on α will represent the coefficient of $R \log \frac{-\Box}{\mu^2} R$ and β the coefficient of $R_{\mu\nu} \log \frac{-\Box}{\mu^2} R^{\mu\nu}$. Observe, notwithstanding, that the last term in (4.2) provides independent contributions in the non-linear regime and, particularly, the background equations of motion (left-hand side of (4.20)) change, but none of this affects the right-hand side of (4.20).

The equations of motion (EOM) corresponding to the quantum action (4.2) is given by

$$G_{\mu\nu} + \Delta G^{L}_{\mu\nu} + \Delta G^{NL}_{\mu\nu} = 8\pi G T_{\mu\nu}, \tag{4.8}$$

where $\Delta G^{L}_{\mu\nu}$ represents the local contribution to the modification of Einstein's tensor and $\Delta G^{NL}_{\mu\nu} = \Delta G^{\alpha}_{\mu\nu} + \Delta G^{\beta}_{\mu\nu}$ is the non-local one (due to the log operator), obtained from the terms proportional to α and β, represented by $\Delta G^{\alpha}_{\mu\nu}$ and $\Delta G^{\beta}_{\mu\nu}$, respectively. Here we shall show only the calculation for the non-local part $\Delta G^{NL}_{\mu\nu}$ as the local contribution can be easily deduced from it. Nevertheless, our final results will be entirely general, including both local and non-local physics. The $\Delta G^{\alpha}_{\mu\nu}$ has been determined in the literature [12]:

$$-\xi \Delta G^{\alpha}_{\mu\nu} = 2\left(R_{\mu\nu} - \frac{1}{4} g_{\mu\nu} R\right)\left(\log \frac{-\Box}{\mu^2} R\right) - 2\left(\nabla_\mu \nabla_\nu - g_{\mu\nu}\Box\right)\left(\log \frac{-\Box}{\mu^2} R\right), \tag{4.9}$$

where $\xi = \frac{1}{16\pi G\alpha}$. Observe that the integral term appearing in [12], which comes from the variation of the D'Alembert operator, is not present here. The reason is that in the weak field limit the variation of the D'Alembert operator leads to curvature terms higher than second order, thus they are safely negligible [13]. The other contribution to $\Delta G_{\mu\nu}$ reads

$$\begin{aligned} \zeta \Delta G^{\beta}_{\mu\nu} &= -\frac{1}{2} g_{\mu\nu} R_{\rho\sigma} \log\left(\frac{-\Box}{\mu^2}\right) R^{\rho\sigma} + \Box \log\left(\frac{-\Box}{\mu^2}\right) R_{\mu\nu} + g_{\mu\nu} \nabla_\rho \nabla_\sigma \log\left(\frac{-\Box}{\mu^2}\right) R^{\rho\sigma} \\ &\quad + R^{\sigma}_{\mu} \log\left(\frac{-\Box}{\mu^2}\right) R_{\nu\sigma} + R^{\sigma}_{\nu} \log\left(\frac{-\Box}{\mu^2}\right) R_{\mu\sigma} \\ &\quad - \nabla_\rho \nabla_\mu \log\left(\frac{-\Box}{\mu^2}\right) R^{\rho}_{\nu} - \nabla_\rho \nabla_\nu \log\left(\frac{-\Box}{\mu^2}\right) R^{\rho}_{\mu} \end{aligned} \tag{4.10}$$

where $\zeta = \frac{1}{16\pi G\beta}$.

4.3 Gravitational Wave Backreaction

The first step is to isolate the fluctuations $h_{\mu\nu}$ (GWs) from the background geometry $\bar{g}_{\mu\nu}$, by splitting the metric as $g_{\mu\nu} = \bar{g}_{\mu\nu} + h_{\mu\nu}$. This separation is only valid in the regime where the GW wavelength λ is much smaller than the background radius L, i.e. $\lambda \ll L$, such that a clear distinction between background and GW can be made. As a first approximation, the background metric $\bar{g}_{\mu\nu}$ will be employed to raise/lower indices as well as to define all the operators, e.g. $\Box = \bar{g}^{\mu\nu}\nabla_\mu\nabla_\nu$. The connection is also assumed to be compatible with $\bar{g}_{\mu\nu}$ instead of $g_{\mu\nu}$.

The separation of gravity into background and fluctuations allows one to expand the Ricci tensor as

$$R_{\mu\nu} = \bar{R}_{\mu\nu} + R^{(1)}_{\mu\nu} + R^{(2)}_{\mu\nu} + O(h^3), \tag{4.11}$$

where the objects with bar are calculated with respect to the background and the rest depends solely on the fluctuation. The superscript (n) denotes the order in h of the underlying quantity. Naively, one could think that the EOM could be calculated order by order in h, producing no backreaction into the background. The caveat is that there are two small parameters in play, i.e. the fluctuation amplitude h and $\varepsilon \equiv \frac{\lambda}{L}$, thus one can compensate the other. Their relation is determined by the EOM[1] and in the presence of external matter

$$h \ll \varepsilon \ll 1, \tag{4.12}$$

as can be seen from Eq. (4.8).

To get the GW backreaction, we need to calculate the average of tensor fields over a region of length scale d, where $\lambda \ll d \ll L$. This forces the high-frequency modes to be eliminated because of their quick oscillation, keeping the low-frequency modes intact. The subtle point is that there is no standard way of summing tensors based on distinct points of a manifold. Isaacson's definition [14, 15] for the average of a tensor comes to our rescue. It is based on the idea of parallel transporting the tensor field along geodesics from each spacetime point to a common position where its different values can be compared:

$$\langle A_{\mu\nu}(x)\rangle = \int j^{\alpha'}_{\mu}(x,x')j^{\beta'}_{\nu}(x,x')A_{\alpha'\beta'}(x')f(x,x')\sqrt{-\bar{g}(x')}d^4x', \tag{4.13}$$

where $j^{\alpha'}_{\mu}$ is the bivector of geodesic parallel displacement and $f(x,x')$ is a weight function that falls quickly and smoothly to zero when $|x-x'| > d$, such that

$$\int_{\text{all space}} f(x,x')\sqrt{-\bar{g}(x')}d^4x' = 1. \tag{4.14}$$

From this definition, the following rules can be proven [2]:

- The average of an odd product of short-wavelength quantities is zero.
- The derivative of a short-wavelength tensor averages to zero, e.g., $\langle\nabla_\mu T^\mu_{\alpha\beta}\rangle = 0$.
- As a corollary, integration by parts can be performed and one can flip derivatives, e.g., $\langle R^\mu_\alpha \nabla_\mu S_\beta\rangle = -\langle S_\beta \nabla_\mu R^\mu_\alpha\rangle$.

Hence, to get the backreaction one has to calculate

[1]Observe that $\bar{R}_{\mu\nu} \sim \frac{1}{L^2}$, $R^{(n)}_{\mu\nu} \sim \frac{h^n}{\lambda^2}$ and the contribution of GWs to the curvature is negligible compared to the contribution of matter sources.

$$\langle G_{\mu\nu}\rangle + \langle \Delta G^{NL}_{\mu\nu}\rangle = 8\pi G\langle T_{\mu\nu}\rangle \tag{4.15}$$

to second order in h (higher orders are utterly small).[2] Taking the average of Eq. (4.9), yields

$$\begin{aligned}-\xi\langle \Delta G^{\alpha}_{\mu\nu}\rangle = 2\left(\left\langle R_{\mu\nu}\log\left(\frac{-\Box}{\mu^2}\right)R\right\rangle - \frac{1}{4}\bar{g}_{\mu\nu}\left\langle R\log\left(\frac{-\Box}{\mu^2}\right)R\right\rangle\right)\\ -2\left\langle(\nabla_\mu\nabla_\nu - g_{\mu\nu}\Box)\log\left(\frac{-\Box}{\mu^2}\right)R\right\rangle.\end{aligned} \tag{4.16}$$

It follows from the above rules that

$$\left\langle R_{\mu\nu}\log\left(\frac{-\Box}{\mu^2}\right)R^{\mu\nu}\right\rangle = \bar{R}_{\mu\nu}\log\left(\frac{-\Box}{\mu^2}\right)\bar{R}^{\mu\nu} + \left\langle R^{(1)}_{\mu\nu}\log\left(\frac{-\Box}{\mu^2}\right)R^{(1)\mu\nu}\right\rangle, \tag{4.17}$$

since the average of linear terms in h is zero. Cross terms (e.g. $\bar{R}R^{(2)}$) do not appear because they are negligible as a result of the condition (4.12). Futhermore, the last line of Eq. (4.16) has a global derivative, thus the high-frequency contribution averages to zero.

Combining the Eqs. (4.16) and (4.17) leads to

$$\begin{aligned}-\xi\langle \Delta G^{\alpha}_{\mu\nu}\rangle = 2\left(\bar{R}_{\mu\nu} - \frac{1}{4}\bar{g}_{\mu\nu}\bar{R}\right)\log\left(\frac{-\Box}{\mu^2}\right)\bar{R} + 2\left(\left\langle R^{(1)}_{\mu\nu}\log\left(\frac{-\Box}{\mu^2}\right)R^{(1)}\right\rangle\right.\\ \left. - \frac{1}{4}\bar{g}_{\mu\nu}\left\langle R^{(1)}\log\left(\frac{-\Box}{\mu^2}\right)R^{(1)}\right\rangle\right)\\ -2(\nabla_\mu\nabla_\nu - \bar{g}_{\mu\nu}\Box)\log\left(\frac{-\Box}{\mu^2}\right)\bar{R}.\end{aligned} \tag{4.18}$$

Analogously, taking the average of Eq. (4.10) yields

$$\begin{aligned}\zeta\langle \Delta G^{\beta}_{\mu\nu}\rangle = -\frac{1}{2}\bar{g}_{\mu\nu}\left(\bar{R}_{\rho\sigma}\log\left(\frac{-\Box}{\mu^2}\right)\bar{R}^{\rho\sigma} + \left\langle R^{(1)}_{\rho\sigma}\log\left(\frac{-\Box}{\mu^2}\right)R^{(1)\rho\sigma}\right\rangle\right)\\ +\Box\log\left(\frac{-\Box}{\mu^2}\right)\bar{R}_{\mu\nu} + \bar{g}_{\mu\nu}\nabla_\rho\nabla_\sigma\log\left(\frac{-\Box}{\mu^2}\right)\bar{R}^{\rho\sigma}\\ +\bar{R}^{\sigma}_{\mu}\log\left(\frac{-\Box}{\mu^2}\right)\bar{R}_{\nu\sigma} + \bar{R}^{\sigma}_{\nu}\log\left(\frac{-\Box}{\mu^2}\right)\bar{R}_{\mu\sigma} + 2\left\langle R^{(1)\sigma}_{\mu}\log\left(\frac{-\Box}{\mu^2}\right)R^{(1)}_{\nu\sigma}\right\rangle\\ -\nabla_\rho\nabla_\mu\log\left(\frac{-\Box}{\mu^2}\right)\bar{R}^{\rho}_{\nu} - \nabla_\rho\nabla_\nu\log\left(\frac{-\Box}{\mu^2}\right)\bar{R}^{\rho}_{\mu}.\end{aligned} \tag{4.19}$$

---

[2] When performing the scalar-vector-tensor decomposition to second order in perturbation theory, one has to take into account the contributions from the coupling between scalar and tensor perturbations [16]. These contributions are automatically being taken into account here as we are not decomposing the metric perturbation and everything is given in terms of the entire perturbation $h_{\mu\nu}$.

Combining Eqs. (4.15), (4.18) and (4.19) gives the background EOM

$$\begin{aligned}
&\bar{R}_{\mu\nu} - \frac{1}{2}\bar{g}_{\mu\nu}\bar{R} - \frac{2}{\xi}\left[\left(\bar{R}_{\mu\nu} - \frac{1}{4}\bar{g}_{\mu\nu}\bar{R}\right)\log\left(\frac{-\Box}{\mu^2}\right)\bar{R} - (\nabla_\mu\nabla_\nu - \bar{g}_{\mu\nu}\Box)\log\left(\frac{-\Box}{\mu^2}\right)\bar{R}\right] \\
&- \frac{1}{2\zeta}\bar{g}_{\mu\nu}\bar{R}_{\rho\sigma}\log\left(\frac{-\Box}{\mu^2}\right)\bar{R}^{\rho\sigma} + \frac{1}{\zeta}\bar{R}^{\sigma}_{\mu}\log\left(\frac{-\Box}{\mu^2}\right)\bar{R}_{\nu\sigma} + \frac{1}{\zeta}\bar{R}^{\sigma}_{\nu}\log\left(\frac{-\Box}{\mu^2}\right)\bar{R}_{\mu\sigma} \\
&+ \frac{1}{\zeta}\Box\log\left(\frac{-\Box}{\mu^2}\right)\bar{R}_{\mu\nu} + \frac{1}{\zeta}\bar{g}_{\mu\nu}\nabla_\rho\nabla_\sigma\log\left(\frac{-\Box}{\mu^2}\right)\bar{R}^{\rho\sigma} - \frac{1}{\zeta}\nabla_\rho\nabla_\mu\log\left(\frac{-\Box}{\mu^2}\right)\bar{R}^{\rho}_{\nu} \\
&- \frac{1}{\zeta}\nabla_\rho\nabla_\nu\log\left(\frac{-\Box}{\mu^2}\right)\bar{R}^{\rho}_{\mu} \\
&= 8\pi G(\langle T_{\mu\nu}\rangle + t^{GR}_{\mu\nu} + t^{NL}_{\mu\nu}),
\end{aligned} \tag{4.20}$$

where $t^{GR}_{\mu\nu}$ is the classical contribution to the GW energy-momentum tensor:

$$t^{GR}_{\mu\nu} = -\frac{1}{8\pi G}\left(\langle R^{(2)}_{\mu\nu}\rangle - \frac{1}{2}\bar{g}_{\mu\nu}\langle R^{(2)}\rangle\right) \tag{4.21}$$

and $t^{NL}_{\mu\nu}$ is the non-local one:

$$\begin{aligned}
t^{NL}_{\mu\nu} = -\frac{1}{8\pi G}\Bigg[&-\frac{2}{\xi}\left(\langle R^{(1)}_{\mu\nu}\log\left(\frac{-\Box}{\mu^2}\right)R^{(1)}\rangle - \frac{1}{4}\bar{g}_{\mu\nu}\langle R^{(1)}\log\left(\frac{-\Box}{\mu^2}\right)R^{(1)}\rangle\right) \\
&+ \frac{2}{\zeta}\langle R^{(1)\sigma}_{\mu}\log\left(\frac{-\Box}{\mu^2}\right)R^{(1)}_{\nu\sigma}\rangle - \frac{1}{2\zeta}\bar{g}_{\mu\nu}\langle R^{(1)}_{\rho\sigma}\log\left(\frac{-\Box}{\mu^2}\right)R^{(1)\rho\sigma}\rangle\Bigg].
\end{aligned} \tag{4.22}$$

In the same way, the local contribution is given by

$$\begin{aligned}
t^{L}_{\mu\nu} = -\frac{1}{8\pi G}\Bigg[&-32\pi G b_1\left(\langle R^{(1)}_{\mu\nu}R^{(1)}\rangle - \frac{1}{4}\bar{g}_{\mu\nu}\langle R^{(1)}R^{(1)}\rangle\right) \\
&+ 32\pi G b_2\langle R^{(1)\sigma}_{\mu}R^{(1)}_{\nu\sigma}\rangle - 8\pi G b_2\bar{g}_{\mu\nu}\langle R^{(1)}_{\rho\sigma}R^{(1)\rho\sigma}\rangle\Bigg].
\end{aligned} \tag{4.23}$$

Hence, the total GW energy-momentum tensor is $t_{\mu\nu} = t^{GR}_{\mu\nu} + t^{L}_{\mu\nu} + t^{NL}_{\mu\nu}$.

At this point, some comments are needed. First of all, note that the left-hand side of Eq. (4.20) corresponds only to the background effect, which we interpret as pure gravity. In fact, the left-hand side is precisely the same as in Eq. (4.8) when $\bar{g}_{\mu\nu}$ takes the place of $g_{\mu\nu}$. The right-hand side constitutes the matter sector, as usual, but with the addition of the GW contribution. This contribution is the most general energy-momentum tensor to leading order, accounting for both classical and quantum effects. Observe that, because of the diffeomorphism invariance of the model, the total energy-momentum tensor is covariantly conserved

$$\nabla^\mu(T_{\mu\nu} + t_{\mu\nu}) = 0, \tag{4.24}$$

which express the energy and momentum exchange between matter sources and GWs. Far away from the source, this leads to the conservation of the GW energy-momentum tensor

$$\partial^{\mu} t_{\mu\nu} = 0. \tag{4.25}$$

So far, no gauge conditions have been used and $t_{\mu\nu}$ also accounts for non-physical degrees of freedom. To remove them, we will take the limit where the GW is far away from the source, such that the background is almost flat and the linear EOM becomes [11]

$$\Box_{\eta} h_{\mu\nu} + 16\pi G \left[b_2 + \beta \log\left(\frac{-\Box_{\eta}}{\mu^2} \right) \right] \Box_{\eta}^2 h_{\mu\nu} = 0, \tag{4.26}$$

where $\Box_{\eta} = \eta^{\mu\nu}\partial_{\mu}\partial_{\nu}$ is the flat D'Alembert operator. Observe that the parameter α is not present in Eq. (4.26). This happens because α is proportional to terms depending on the trace h, which can be taken to be zero far away from the source. Using the gauge $\partial^{\nu} h_{\mu\nu} = 0$ and $h = 0$ (only meaningful outside the source) together with Eq. (4.26) in the definition of $t_{\mu\nu}$ leads to

$$t_{\mu\nu} = \frac{1}{8\pi G} \left[\frac{1}{4} \langle \partial_{\mu} h_{\alpha\beta} \partial_{\nu} h^{\alpha\beta} \rangle + \frac{1}{2} \langle h_{\mu}^{\sigma} \Box_{\eta} h_{\nu\sigma} \rangle - \frac{1}{8} \eta_{\mu\nu} \langle h_{\rho\sigma} \Box_{\eta} h^{\rho\sigma} \rangle \right]. \tag{4.27}$$

Comparing this result to Eq. (4.1), we can clearly see that the first term in $t_{\mu\nu}$ corresponds to classical GR, while the other two come from quantum corrections. Note that the log operators do not show up explicitly in Eq. (4.27) as the gravitational field is on shell. This means that their contribution will just appear in the field solutions. For the same reason, the procedure (4.3) of imposing causality does not need to be followed at this stage since the non-local effects will only be reflected in the solutions for $h_{\mu\nu}$. The parameters b_2 and β now only show up in the mass m of $h_{\mu\nu}$.

Hence, the GW energy density reads

$$\rho \equiv t_{00} = \frac{1}{8\pi G} \left[\frac{1}{4} \langle \dot{h}_{\alpha\beta} \dot{h}^{\alpha\beta} \rangle + \frac{1}{2} \langle h_0^{\alpha} \Box_{\eta} h_{0\alpha} \rangle + \frac{1}{8} \langle h_{\rho\sigma} \Box_{\eta} h^{\rho\sigma} \rangle \right]. \tag{4.28}$$

As a concrete example, let us take a plane wave solution propagating in the z direction

$$h_{\mu\nu} = \epsilon_{\mu\nu} \cos(\omega t - kz). \tag{4.29}$$

Substituting this into Eq. (4.28) yields

$$\rho = \frac{1}{16\pi G} \left[\frac{\epsilon^2 \omega^2}{4} + \frac{1}{2} \left(\epsilon_0^{\alpha} \epsilon_{0\alpha} + \frac{\epsilon^2}{4} \right) (\omega^2 - k^2) \right], \tag{4.30}$$

where $\epsilon^2 = \epsilon_{\mu\nu}\epsilon^{\mu\nu}$. Hence, modifications in the dispersion relation lead to observable differences of the GW energy. In the case of a classical wave, i.e. $\omega^2 = k^2$, the second

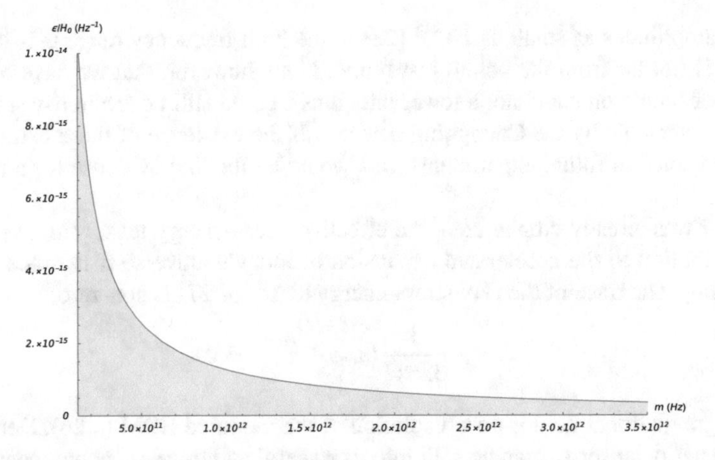

Fig. 4.1 The blue area in the graph represents the allowed region of the parameter space (m, ϵ)

term vanishes identically, which results in the classical energy as one would expect. In the most general case, there could be complex frequencies leading to damping as argued in [17–20]. In such case, Eq. (4.30) can be easily generalized. Observe that the second term in (4.30) is proportional to the particle's mass m, thus it is constant as any change in the frequency would be balanced by a change in the momentum. Dividing the constant term by the critical density $\rho_c = \frac{3H_0^2}{8\pi G}$, where H_0 is the today's Hubble constant, yields the frequency-independent gravitational wave density parameter Ω_0, which has been limited to be smaller than 1.7×10^{-7} by LIGO [21]:

$$\Omega_0 = \frac{1}{12}\left(\epsilon_0^{\alpha}\epsilon_{0\alpha} + \frac{\epsilon^2}{4}\right)\frac{m^2}{H_0^2} < 1.7 \times 10^{-7}. \tag{4.31}$$

We remind the reader that the initial parameters b_2 and β only show up in terms of the mass m as the field $h_{\mu\nu}$ is on shell. Figure 4.1 illustrates the permitted region of the parameter space (m, ϵ). By using the lower bound on the mass of the complex pole[3] obtained in [17], i.e. $m > 5 \times 10^{-13}$GeV, allows us to translate the above bound to

$$\epsilon < 1.4 \times 10^{-33}. \tag{4.32}$$

To the best of our knowledge, this is the first bound ever obtained on the amplitude of the massive mode. It is 12 orders of magnitude smaller than the strain sensibility of LIGO's interferometer, which can probe amplitudes up to $\sim 10^{-22}$ in the frequency range from 10 Hz to 10 kHz. Although it seems hopeless to reach such small distances, the Chongqing University detector (currently under construction) will be able to

[3]This conservative bound, and consequently the bound on ϵ, was found assuming all the energy of a merger goes into the complex mode. Naturally, this does not represent the real situation as the classical mode should also be excited. In a more detailed analysis, we expect to obtain a better bound.

probe amplitudes as small as 10^{-32} [22] in the high-frequency range 0.1–10 GHz, which is not far from the bound just found. Note, however, that we have obtained an upper bound on ϵ and not a lower one, thus ϵ could still be arbitrarily small and not be detectable by the Chongqing detector. If the existence of these extra modes are confirmed in future experiments, this would be the first evidence for a massive mode.

As it was already emphasized, the effective stress-energy tensor might produce a contribution to the accelerated expansion of today's universe if its trace is non-vanishing. The trace of the GW stress-energy tensor (4.27) is non-zero:

$$t = -\frac{1}{32\pi G}\langle h_{\alpha\beta}\Box_\eta h^{\alpha\beta}\rangle \neq 0, \tag{4.33}$$

as the gravitational field is now a solution of the modified EOM (4.26). Hence, the stress-energy tensor $t_{\mu\nu}$ can be split into its traceful and traceless components

$$t_{\mu\nu} = t_{\mu\nu} - \frac{1}{4}\eta_{\mu\nu}t + \frac{1}{4}\eta_{\mu\nu}t \tag{4.34}$$

and the cosmological constant can be identified as

$$\Lambda \equiv \frac{1}{16}\langle h_{\alpha\beta}\Box_\eta h^{\alpha\beta}\rangle = \frac{1}{16}\epsilon^2 m^2, \tag{4.35}$$

where in the second equality the plane wave solution (4.29) was used. After taking the average, Λ varies very slowly with space and time. In fact, it is exactly constant across any region of length d and its variation only becomes significant in a region containing many lengths of size d. Hence, for our goals, we can safely neglect the spacetime dependence of the emergent cosmological constant Λ and take it as a constant. Remember that, initially, the bare cosmological constant was set to zero. A non-zero initial or bare cosmological constant Λ_b would just be shifted by the Λ obtained above and the physical cosmological constant would be $\Lambda_{eff} \equiv \Lambda_b + \Lambda$. The major proposition here is that quantum gravitational waves produce a non-vanishing contribution to the cosmological constant Λ_{eff}. In this context, the new gravitational interactions and degrees of freedom showing up in high energies are represented by non-local effects in the IR. The latter, combined with the local interactions, leads to a gravitational stress-energy tensor whose trace is non-zero and which contributes to the total cosmological constant.

4.4 Gravitational Wave Propagation

Only the physics of the low-frequency modes has been considered so far. For completeness, we now turn our attention to the high-frequency waves, which leads to the equation describing the GW propagation in curved spacetime. This can be straigthforwardly done by subtracting the background Eq. (4.20) from the total EOM (4.8)

$$G_{\mu\nu} + \Delta G_{\mu\nu} - \langle G_{\mu\nu} + \Delta G_{\mu\nu} \rangle = 8\pi G(T_{\mu\nu} - \langle T_{\mu\nu} \rangle), \tag{4.36}$$

where $\Delta G_{\mu\nu} = \Delta G^{L}_{\mu\nu} + \Delta G^{NL}_{\mu\nu}$. If we ignore the local terms for a moment and keep only the terms up to linear order in h and λ/L, we obtain

$$R^{(1)}_{\mu\nu} - \frac{1}{2}\bar{g}_{\mu\nu}R^{(1)} + \frac{2}{\xi}(\nabla_\mu\nabla_\nu - \bar{g}_{\mu\nu}\Box)\log\left(\frac{-\Box}{\mu^2}\right)R^{(1)} + \frac{1}{\zeta}\Bigg[\Box\log\left(\frac{-\Box}{\mu^2}\right)R^{(1)}_{\mu\nu}$$
$$+ \bar{g}_{\mu\nu}\nabla_\rho\nabla_\sigma\log\left(\frac{-\Box}{\mu^2}\right)R^{(1)\rho\sigma} - \nabla_\rho\nabla_\mu\log\left(\frac{-\Box}{\mu^2}\right)R^{(1)\rho}_{\nu} - \nabla_\rho\nabla_\nu\log\left(\frac{-\Box}{\mu^2}\right)R^{(1)\rho}_{\mu}\Bigg] = 0 \tag{4.37}$$

Outside the matter source, the gauge $\nabla^\nu h_{\mu\nu} = 0$ can be used together with $h = 0$, which gives

$$\Box h_{\mu\nu} + 16\pi G\beta\log\left(\frac{-\Box}{\mu^2}\right)\Box^2 h_{\mu\nu} = 0. \tag{4.38}$$

Similarly, adding the local curvature terms yields

$$\Box h_{\mu\nu} + 16\pi G\left[b_2 + \beta\log\left(\frac{-\Box}{\mu^2}\right)\right]\Box^2 h_{\mu\nu} = 0. \tag{4.39}$$

When deducing Eq. (4.39), we used the commutation relation of covariant derivatives, discarding terms proportional to the background curvature as they only contribute to higher orders in λ/L. Equation (4.39) describes the propagation of GWs in an arbitrary curved background in the absence of external matter, when the only source for a non-zero curvature is the GW stress-energy tensor. The curvature terms do not show up as they give no contribution to leading order. We conclude that the equation for the propagation of GWs in curved spacetimes can be found by performing a simple "minimal coupling" recipe to Eq. (4.26) where spacetime is flat, that is, by making the following replacement

$$\eta_{\mu\nu} \to \bar{g}_{\mu\nu}, \tag{4.40}$$
$$\partial_\mu \to \nabla_\mu. \tag{4.41}$$

Equations (4.20) and (4.39) together describe the entire classical and quantum process (to leading order) of the GW self-gravitation: small fluctuations around spacetime change the curvature, which in turn modify the GW's trajectory and vice-versa.

4.5 Conclusions

We showed in this chapter how to evaluate the quantum corrections to the GW stress-energy tensor. Quantum effects promote the traceless tensor $t_{\mu\nu}^{\mathrm{GR}}$ to a traceful quantity that contributes to the accelerated expansion of the present universe. Futhermore, the energy density component receives a dependence on modifications to the dispersion relation, proving itself to be a useful quantity to probe when looking for quantum gravitational effects in the laboratory. In fact, by using the latest LIGO's data, it was found a new upper bound on the amplitude of the massive mode. We also showed that the high-frequency wave equation led to a generalization of the wave equation (4.26) to arbitrary curved spacetimes (4.39). Such a generalization is important to the investigation of quantum GW solutions in cosmology and in the early universe where the spacetime is highly curved. Lastly, it must be emphasized once again that these quantum contributions are model independent (since they are obtained using Effective Field Theory), thus constituting genuine predictions of Quantum Gravity, which gives us some hints of how a complete theory for quantum gravity should behave below the Planck scale, should such a theory exist.

References

1. Abbott BP et al (2016) Observation of gravitational waves from a binary black hole merger. Phys Rev Lett 116(6):061102
2. Stein LC, Yunes N (2011) Effective gravitational wave stress-energy tensor in alternative theories of gravity. Phys Rev D 83:064038
3. Preston AWH (2016) Cosmological backreaction in higher-derivative gravity expansions. JCAP 1608(08):038
4. Preston AWH, Morris TR (2014) Cosmological back-reaction in modified gravity and its implications for dark energy. JCAP 1409:017
5. Saito K, Ishibashi A (2013) High frequency limit for gravitational perturbations of cosmological models in modified gravity theories. PTEP 2013:013E04
6. Berry CPL, Gair JR (2011) Linearized f(R) gravity: gravitational radiation and solar system tests. Phys Rev D 83:104022. (Erratum: Phys Rev D 85:089906, 2012)
7. Rasanen S (2010) Backreaction as an alternative to dark energy and modified gravity
8. Rasanen S (2004) Dark energy from backreaction. JCAP 0402:003
9. Buchert T, Rasanen S (2012) Backreaction in late-time cosmology. Ann Rev Nucl Part Sci 62:57–79
10. Donoghue JF, El-Menoufi BK (2014) Nonlocal quantum effects in cosmology: quantum memory, nonlocal FLRW equations, and singularity avoidance. Phys Rev D 89(10):104062
11. Calmet X, Capozziello S, Pryer D (2017) Gravitational effective action at second order in curvature and gravitational waves. Eur Phys J C 77(9):589
12. Codello A, Jain RK (2017) On the covariant formalism of the effective field theory of gravity and its cosmological implications. Class Quant Grav 34(3):035015
13. Donoghue JF, El-Menoufi BK (2015) Covariant non-local action for massless QED and the curvature expansion. JHEP 10:044
14. Isaacson RA (1968) Gravitational radiation in the limit of high frequency. II. Nonlinear terms and the effective stress tensor. Phys Rev 166:1272–1279
15. Isaacson RA (1967) Gravitational radiation in the limit of high frequency. I. The linear approximation and geometrical optics. Phys Rev 166:1263–1271

16. Marozzi G, Vacca GP (2014) Tensor mode backreaction during slow-roll inflation. Phys Rev D 90(4):043532
17. Calmet X, Kuntz I, Mohapatra S (2016) Gravitational waves in effective quantum gravity. Eur Phys J C 76(8):425
18. Calmet X (2014) The lightest of black holes. Mod Phys Lett A 29(38):1450204
19. Calmet X, Kuntz I (2016) Higgs Starobinsky inflation. Eur Phys J C 76(5):289
20. Calmet X, Casadio R, Kamenshchik AYu, Teryaev OV (2017b) Graviton propagator, renormalization scale and black-hole like states. Phys Lett B 774:332–337
21. Abbott BP et al (2017) Upper limits on the stochastic gravitational-wave background from advanced LIGO's first observing run. Phys Rev Lett 118(12):121101. (Erratum: Phys Rev Lett 119(2) 029901, 2017)
22. Baker RML (2009) The peoples Republic of China high-frequency gravitational wave research program. AIP Conf Proc 1103:548–552

Chapter 5
Higgs-Starobinsky Inflation

In this chapter, we argue that Starobinky inflation could be induced by quantum gravitational effects due to a large non-minimal coupling of the Higgs field to the curvature. The Higgs Starobinsky model solves issues associated to large Higgs boson values in the early universe, which in a metastable universe would not be a correct description of nature. We verify explicitly that these quantum gravitational corrections do not destabilize Starobinsky's potential.

5.1 Introduction

The idea that inflation might be due to fields already present in the standard model of particle physics or quantum general relativity is very attractive and has received a lot of attention in the recent years. Two theories are particularly noticeable because of their simplicity and elegance: Higgs inflation [1–3], which is characterized by a large non-minimal coupling of the Higgs field H to the curvature ($\xi H^{\dagger} HR$) and Starobinsky inflation [4], which is based on the higher order term R^2. They are both simple and in very good agreement with the latest Planck data [5].

These two theories should not be seen as physics beyond the standard model since both operators $\xi H^{\dagger} HR$ and R^2 are expected to be appear when general relativity is coupled to the standard model. The goal of this chapter is to show an interesting and distinct possibility: Starobinsky inflation can be generated by quantum gravitational effects due to a large non-minimal coupling of the Higgs to the curvature. In this formalism, we need not assume that the Higgs field starts off at high values in the early universe, which could relax constraints that comes from the requirement of having a stable Higgs potential even for large Higgs field values [6–8].

© Springer Nature Switzerland AG 2019

I. Kuntz, *Gravitational Theories Beyond General Relativity*,
Springer Theses, https://doi.org/10.1007/978-3-030-21197-4_5

5.2 Effective Field Theory of Gravity

Now we will point out that both terms necessary for Higgs and Starobinsky inflation naturally arises when the standard model is coupled to general relativity. Although a complete understanding of quantum gravity is still an open question, we can use effective field theory techniques to study quantum gravity below the energy scale $M_\star$, supposedly of the order of the Planck mass $M_P = \sqrt{8\pi G_N}^{-1} = 2.4335 \times 10^{18}$ GeV, at which quantum gravitational effects are expected to become large and the effective theory breaks down. At energies below $M_\star$, one can describe all of particle physics and cosmology with the following action (see e.g. [9–11])

$$S = \int d^4x \sqrt{-g}\left(\left(\frac{1}{2}M^2 + \xi H^\dagger H\right) R - \Lambda_C^4 + c_1 R^2 + c_2 C^2 + c_3 E + c_4 \Box R \right.$$
$$\left. - L_{SM} - L_{DM} + O(M_\star^{-2})\right) \tag{5.1}$$

where we are including only operators of dimension four, which are expected to dominate in the infrared. Observe that we are using the Weyl basis and the following notations: R is the Ricci scalar, $R^{\mu\nu}$ is the Ricci tensor, $E = R_{\mu\nu\rho\sigma}R^{\mu\nu\rho\sigma} - 4R_{\mu\nu}R^{\mu\nu} + R^2$, $C^2 = E + 2R_{\mu\nu}R^{\mu\nu} - 2/3R^2$, the dimensionless ξ stands for the non-minimal coupling of the Higgs field H to the curvature, the coefficients c_i are dimensionless free parameters, the cosmological constant Λ_C is of order of 10^{-3} eV, the Higgs boson vacuum expectation value $v = 246$ GeV contributes to the value of the Planck scale

$$(M^2 + \xi v^2) = M_P^2, \tag{5.2}$$

L_{SM} has all the standard model interactions and, at last, L_{DM} describes the dark matter sector, which is the only sector that has not been experimentally tested yet. Submillimeter pendulum experiments of Newton's law [12] lead to very weak bounds on the coefficients c_i. In the absence of accidental cancellations between these parameters, they are constrained to be smaller than 10^{61} [13]. The discovery of the Higgs field and precision measurements of its couplings to fermions and bosons at the LHC can be used to find a bound on ξ, which is given by $|\xi| < 2.6 \times 10^{15}$ [14]. Evidently, very little is known about the values of c_i and ξ.

5.3 Inflation in Quantum Gravity

In addition to describing the standard model and gravity, the action Eq. (5.1) can also describe inflation when some of its parameters take specific values and if some of its fields fulfill specific initial conditions in the early universe. This action, depending on the initial conditions, can describe either Higgs inflation when $\xi \sim 10^4$ and the

Higgs field is chosen to take large values in the early universe or Starobinsky inflation when $c_1 \sim 10^9$ and the scalaron, which can be made explicit in the Einstein frame, takes large values in the early universe.

Assuming that the Higgs field take small values in the early universe, Eq. (5.1) becomes

$$S^J_{Starobinsky} = \int d^4x\sqrt{g}\frac{1}{2}\left(M_P^2 R + c_S R^2\right) \quad (5.3)$$

during inflation, which in the Einstein frame it looks like

$$S^E_{Starobinsky} = \int d^4x\sqrt{g}\left(\frac{M_P^2}{2}R - \frac{1}{2}\partial_\mu\sigma\partial^\mu\sigma - \frac{M_P^4}{c_S}\left(1 - \exp\left(-\sqrt{\frac{2}{3}}\frac{\sigma}{M_P}\right)\right)^2\right). \quad (5.4)$$

We are assuming that the scalaron σ initially hidden in R^2 is large in the early universe. A successful prediction of the density perturbation $\delta\rho/\rho$ requires $c_S = 0.97 \times 10^9$ [15, 16]. On the other hand, assuming that is the Higgs boson who is large in the early universe, the action (5.1) becomes

$$\begin{aligned} S^J_{Higgs} &= \int d^4x\sqrt{-g}\left(\frac{M^2}{2}R + \xi_H H^\dagger H R - L_{SM}\right) \quad (5.5) \\ &= \int d^4x\sqrt{-g}\left(\frac{M^2 + \xi_H h^2}{2}R - \frac{1}{2}\partial_\mu h\partial^\mu h + \frac{\lambda}{4}(h^2 - v^2)^2\right) + \cdots. \end{aligned}$$

Transforming to the Einstein frame, we get

$$S^E_{Higgs} = \int d^4x\sqrt{\hat{g}}\left(\frac{M_P^2}{2}\hat{R} - \frac{1}{2}\partial_\mu\chi\partial^\mu\chi + U(\chi) + \cdots\right) \quad (5.6)$$

with

$$\frac{d\chi}{dh} = \sqrt{\frac{\Omega^2 + 6\xi_H^2 h^2/M_P^2}{\Omega^4}} \quad (5.7)$$

where $\Omega^2 = 1 + \xi_H^2 h^2/M_P^2$ and

$$U(\chi) = \frac{1}{\Omega(\chi)^4}\frac{\lambda}{4}(h(\chi)^2 - v^2)^2. \quad (5.8)$$

Againg using the prediction of the density perturbation $\delta\rho/\rho$, inflation requires $\xi_H = 1.8 \times 10^4$.

5.4 Higgs-Starobinsky Inflation

Both theories are attractive as they do not require physics beyond the standard model. Moreover, they both agree with the latest cosmological observations, which favours small tensor perturbations that have not been seen so far. It has also been argued that both models are phenomenologically very similar [17, 18]. Nonetheless, while Starobinky inflation does not experience any known issue, the measurement of the Higgs and top quark masses indicate that the vacuum is metastable, possessing a lower minimum at large values of the Higgs field that would cause problems during the slow-roll evolution of the Higgs as we would end up in the vacuum that does not correspond to the universe as we know it. In this chapter, we argue that there is a third option: when quantum corrections are considered, a large non-minimal coupling of the Higgs boson to curvature can produce Starobinsky inflation by giving rise to a high value for the coefficient of R^2 in the early universe. While our model corresponds to the same physics as Starobinsky inflation, the Higgs field plays a fundamental role as it triggers inflation through the generation of a large coefficient for R^2.

After quantization, the action Eq. (5.1) is required to be renormalized. The cosmological constant will be neglected as it is not important for inflation purposes. In this scenario, Newton's constant does not get any correction to leading order. In contrast, the coefficient c_1 of R^2 gets renormalized and it gains a dependence on the energy scale, which is described by its renormalization group equation. Considering N_s scalar fields non-minimally coupled to the curvature, we obtain the following renormalization group equation to leading order [9–11]

$$\mu \partial_\mu c_1(\mu) = \frac{(1-12\xi)^2}{1152\pi^2} N_s. \tag{5.9}$$

Note that we are neglecting the graviton contribution, which is suppressed by $1/\xi$. Fermions and vector fields also do not contribute to the renormalization of c_1 in the Weyl basis. The renormalization group equation can be straightforwardly integrated, we obtain [9–11]:

$$c_1(\mu_2) = c_1(\mu_1) + \frac{(1-12\xi)^2 N_s}{1152\pi^2} \log \frac{\mu_2}{\mu_1}. \tag{5.10}$$

The bound on c_1 in today's universe are extremely weak as we mentioned before. Even if $c_1(today)$ is of order unity, it would have been large in the early universe if the Higgs non-minimal coupling ξ is large. In fact, assuming that inflation took place at the energy scale $\mu \sim 10^{15}$ GeV, the log term is a factor of order 60 if we take the scale μ_1 of the order of the cosmological constant. A non-minimal coupling of the Higgs to the curvature of $\xi = 1.8 \times 10^4$ would then lead to a coefficient $c_1 = 0.97 \times 10^9$ for R^2. Assuming also that the scalaron from R^2 started off at high field values, we conclude that a large non-minimal coupling of the Higgs field to the curvature can trigger Starobinsky inflation even if the standard model vacuum is metastable as the

Higgs field itself does not roll down its potential during inflation. Inflation is due entirely to the R^2 but is triggered by the Higgs large non-minimal coupling.

Let us stress two fundamental points. The first one is that $c_1 \sim 0.97 \times 10^9$ is fixed by the CMB constraint [15]. This parameter is only this large at inflationary scales because of its renormalization group equation. The second one is that we are neglecting the running of the Higgs boson non-minimal coupling to the curvature as it does not contribute to leading order. The leading contributions of the standard model to the beta-function of the non-minimal coupling are given by [19]:

$$\beta_\xi = \frac{6\xi + 1}{(4\pi)^2}\left[2\lambda + y_t^2 - \frac{3}{2}g^2 - \frac{1}{4}g'^2\right] \tag{5.11}$$

where λ is the self-interaction coupling of the Higgs boson, y_t is the top quark Yukawa coupling, g the SU(2) gauge coupling and g' the U(1) gauge coupling. Quantum gravitational corrections will be suppressed by powers of the Planck mass and can thus be safely ignored as long as we are at energies below the Planck mass.

5.5 Stability Under Radiative Corrections

One worry concerns the fact that if the large non-minimal coupling of the Higgs boson is able to generate a large coefficient for R^2, it could also produce other terms in the effective action that could compromise the shape of the potential and, consequently, the possibility of producing inflation. The leading order quantum effective action up to second order in the curvature expansion that is induced by scalar fields non-minimally coupled to gravity is given by [9, 10]:

$$S_{EFT} = \frac{1}{16\pi G}\int d^4x\sqrt{-g}\left(R + \alpha R^2 + \beta R \log\frac{-\Box}{\mu^2}R + \gamma C^2 + \cdots\right). \tag{5.12}$$

Observe that we are neglecting the cosmological constant, $\alpha = c_1 \times 16\pi G$ and $\gamma = c_2 \times 16\pi G$ are renormalized coupling constants. We also assume that c_2 is small at the scale of inflation as it is not sensitive to the non-minimal coupling of the Higgs to the curvature because we fixed the Higgs non-minimal coupling such that $c_1 = 0.97 \times 10^9$. The coefficient β is a genuine prediction of quantumg gravity and is given by $N_s(1 - 12\xi)^2/(2304\pi^2) \times 16\pi G$ [10] where N_s is the number of scalar degrees of freedom that have been integrated out in the fundamental action, which in our case is 4. The coefficient $N_s(1 - 12\xi)^2/(2304\pi^2)$ is, in fact, large and of the order of 7.8×10^6, thus we must check that the log correction does not produce any sizable effect on the scalaron potential. Before checking this explicitly, we should stress that the large non-minimal coupling between the Higgs boson and the curvature, which is required to generate Starobinsky inflation, does not lead to perturbative unitarity issues [20] (see Appendix A).

Observe that the coefficients of E and of C^2 do not depend on the non-minimal coupling between the Higgs boson and the curvature. Moreover, E does not contribute to the equations of motion in 4 dimensions due to its topological invariance. Before renormalization, the coefficient of C^2 is assumed to be of the same order as that of R^2, i.e. of order 1. Nonetheless, after renormalization, the coefficient of R^2 is tuned to be very large and of the order of 10^9, while the coefficient of C^2 remains comparatively small and is, therefore, negligible.

We will treat the quantum effective action (5.12) as a $F(R)$ gravity with $F(R) = R + \alpha R^2 + \beta R \log \frac{-\Box}{\mu^2} R$. There is a very well known procedure to map such theories from the Jordan frame to the Einstein frame, which simply consists in performing a conformal transformation, see e.g. [21]. The potential for the inflation in the Einstein frame reads

$$V(\phi) = \frac{1}{2\kappa^2}\left(e^{\sqrt{\frac{2}{3}}\kappa\phi} R(\phi) - e^{2\sqrt{\frac{2}{3}}\kappa\phi} F(R(\phi))\right) \tag{5.13}$$

where $\kappa^2 = 8\pi G$ and $R(\phi)$ is a solution to the equation

$$\phi = -\sqrt{\frac{3}{2}}\frac{1}{\kappa}\log\frac{dF(R)}{dR}. \tag{5.14}$$

A formal solution to this equation can be obtained

$$R(\phi) = \frac{1}{2\alpha}\left(\frac{1}{1 + \frac{\beta}{2\alpha}\log\left(\frac{-\Box}{\mu^2}\right)}\right)\left(e^{-\sqrt{\frac{2}{3}}\kappa\phi} - 1\right). \tag{5.15}$$

This expression for $R(\phi)$ can be expanded as a Taylor series in $\frac{\beta}{2\alpha}$, which is in fact a small parameter:

$$R(\phi) = \frac{1}{2\alpha}\left(1 - \sum_{n=1}^{\infty}(-1)^{n+1}\left(\frac{\beta}{2\alpha}\log\left(\frac{-\Box}{\mu^2}\right)\right)^n\right)\left(e^{-\sqrt{\frac{2}{3}}\kappa\phi} - 1\right). \tag{5.16}$$

where the log-term can be represented as

$$\log\left(\frac{-\Box}{\mu^2}\right) = \int_0^{\infty} ds \left(\frac{1}{\mu^2 + s} - \frac{1}{-\Box + s}\right). \tag{5.17}$$

The zeroth order term in $\frac{\beta}{2\alpha} \sim 4 \times 10^{-3}$ is associated to the standard Starobinsky solution:

$$R(\phi)^{(0)} = R(\phi)_{Starobinsky} = \frac{1}{2\alpha}\left(e^{-\sqrt{\frac{2}{3}}\kappa\phi} - 1\right). \tag{5.18}$$

The series expansion produces higher order terms corresponding to operators of the type $\exp(-\sqrt{\frac{2}{3}}\kappa\phi)(2/3\kappa^2\partial_\mu\phi\partial^\mu\phi - \sqrt{2/3}\kappa\Box\phi)$ and higher derivatives thereof. These new terms are however suppressed by powers of $\frac{\beta}{2\alpha}$ and can be safely discarded. It is straightforward to verify that the log-term appearing in the $F(R)$ term of the potential (5.13) is also suppressed by $\frac{\beta}{2\alpha}$ in comparison to the standard scalaron potential.

We thus conclude that the large quantum corrections induced by the large Higgs boson non-minimal coupling do not destabilize the flatness of the scalaron potential. Let us add a few remarks. The theory introduced above is not really a new model. Physics (including reheating or preheating and all of particle physics) is the same as that predicted in Starobinsky inflation. We merely identify a new connection between the Higgs field and inflation. As in the case of the standard Starobinsky model, a coupling $\phi^2 h^2$ will be produced. Nonetheless, it is suppressed by factors of m_{Higgs}^2/M_P^2 which is extremely small. Therefore, particle physics is not affected and the Higgs boson behaves as usual. In addition, the Higgs does not take large field values in the early universe and we can thus safely ignore the term $H^\dagger HR$ when studying the inflationary potential. Observe that there are subtleties when considering the equivalence of quantum corrections in different parameterizations/representations of the model (i.e. when going from the Jordan to the Einstein frame). Here we are avoiding this problem by renormalizing the theory in the Jordan frame, where the model is defined, and only then mapping the effective action to the Einstein frame. When proceeding this way, there are no ambiguities or risk of mixing up the orders in perturbation theory and the expansion in the conformal factor (see e.g. [22–24]).

5.6 Conclusions

In this chapter, we have shown the existence of a interesting connection between the Higgs field and inflation. In the model introduced here, the Higgs field is not assumed to be the inflaton and it is thus at the origin of scalaron potential, but it is required to produce inflation via the generation of a large Wilson coefficient for R^2. This mechanism is interesting as it does not need physics beyond the standard model. The Higgs field does not take large values in the early universe and we could thus be living in a metastable potential.

References

1. Bezrukov FL, Shaposhnikov M (2008) The standard model Higgs boson as the inflaton. Phys Lett B 659:703–706
2. Barvinsky AO, Kamenshchik AYu, Starobinsky AA (2008) Inflation scenario via the standard model Higgs boson and LHC. JCAP 0811:021

3. Barvinsky AO, Kamenshchik AYu, Kiefer C, Starobinsky AA, Steinwachs C (2009) Asymptotic freedom in inflationary cosmology with a non-minimally coupled Higgs field. JCAP 0912:003
4. Starobinsky AA (1980) A new type of isotropic cosmological models without singularity. Phys Lett B91, 99–102 (1980). (771)
5. Akrami Y et al (2018) Planck 2018 results. X. Constraints on inflation
6. Kobakhidze, A, Spencer-Smith A (2014) The Higgs vacuum is unstable
7. Degrassi G, Di Vita S, Elias-Miro J, Espinosa JR, Giudice GF, Isidori G, Strumia A (2012) Higgs mass and vacuum stability in the standard model at NNLO. JHEP 08:098
8. Bezrukov F, Kalmykov MYu, Kniehl BA, Shaposhnikov M (2012) Higgs boson mass and new physics. JHEP **10**, 140 (2012). (275)
9. Codello A, Jain RK (2016) On the covariant formalism of the effective field theory of gravity and leading order corrections. Class Quant Grav 33(22):225006
10. Donoghue JF, El-Menoufi BK (2014) Nonlocal quantum effects in cosmology: quantum memory, nonlocal FLRW equations, and singularity avoidance. Phys Rev D 89(10):104062
11. Birrell ND, Davies PCW (1984) Quantum Fields in Curved Space. Cambridge Monographs on Mathematical Physics. Cambridge University Press, Cambridge
12. Hoyle CD, Kapner DJ, Heckel BR, Adelberger EG, Gundlach JH, Schmidt U, Swanson HE (2004) Sub-millimeter tests of the gravitational inverse-square law. Phys Rev D 70:042004
13. Calmet X, Hsu SDH, Reeb D (2008) Quantum gravity at a TeV and the renormalization of Newton's constant. Phys Rev D 77:125015
14. Atkins M, Calmet X (2013) Bounds on the nonminimal coupling of the Higgs boson to gravity. Phys Rev Lett 110(5):051301
15. Netto TDP, Pelinson AM, Shapiro IL, Starobinsky AA (2016) From stable to unstable anomaly-induced inflation. Eur Phys J C **76**(10), 544 (2016)
16. Starobinsky AA (1983) The perturbation spectrum evolving from a nonsingular initially de-Sitter cosmology and the microwave background anisotropy. Sov Astron Lett 9:302
17. Bezrukov FL, Gorbunov DS (2012) Distinguishing between R^2-inflation and Higgs-inflation. Phys Lett B 713:365–368
18. Salvio A, Mazumdar A (2015) Classical and quantum initial conditions for Higgs inflation. Phys Lett B 750:194–200
19. Buchbinder IL, Odintsov SD, Shapiro IL (1992) Effective action in quantum gravity
20. Calmet X, Casadio R (2014) Self-healing of unitarity in Higgs inflation. Phys Lett B 734:17–20
21. Sebastiani L, Myrzakulov R (2015) F(R) gravity and inflation. Int J Geom Meth Mod Phys 12(9):1530003
22. Calmet X, Yang T-C (2013) Frame transformations of gravitational theories. Int J Mod Phys A 28:1350042
23. Kamenshchik AYu, Steinwachs CF (2015) Question of quantum equivalence between Jordan frame and Einstein frame. Phys Rev D 91(8):084033
24. Vilkovisky GA (1984) The unique effective action in quantum field theory. Nucl Phys B 234:125–137

Chapter 6
Vacuum Stability During Inflation

In the lack of new physics around 10^{10} GeV, the electroweak vacuum is at best metastable. This constitutes an important challenge for large field inflationary models as, during the early rapid evolution of the universe, it seems hard to understand how the Higgs vacuum would not decay into the true lower vacuum of the theory, leading to catastrophic consequences if inflation took place at a scale above 10^{10} GeV. In this chapter, we show that the non-minimal coupling of the Higgs field to the Ricci scalar could solve this issue by producing a direct coupling of the Higgs field to the inflaton potential, thus stabilizing the electroweak vacuum. For specific values of the Higgs field initial condition and of its non-minimal coupling, the inflaton can quickly drive the Higgs field to the electroweak vacuum during inflation.

6.1 Introduction

The non-minimal coupling $\xi\phi^2 R$ of scalar fields (ϕ) to the Ricci scalar R has attracted a lot of attention in the past few years. Indeed, in four space-time dimensions, ξ is a dimensionless coupling constant and as such is likely to be a fundamental constant of nature. With the discovery of the Higgs boson, the only fundamental scalar field found so far, it has become clear that this parameter is important and should be taken into account when coupling the standard model of particle physics to gravity.

The value of the non-minimal coupling of the Higgs field to the Ricci scalar is a free parameter of the standard model. There has been no direct measurement of this fundamental constant of nature yet. The discovery of the Higgs at the Large Hadron Collider at CERN and the fact that the Higgs boson behaves as expected in the standard model suggests that the non-minimal coupling is smaller than 2.6×10^{15} [1]. This constraint comes from the fact that for a high value of the non-minimal coupling, the Higgs field would decouple from the standard model particles. We can

© Springer Nature Switzerland AG 2019

I. Kuntz, *Gravitational Theories Beyond General Relativity*,
Springer Theses, https://doi.org/10.1007/978-3-030-21197-4_6

see that we have very little understanding of the value of this constant. Conformal symmetry would require $\xi = 1/6$, but this invariance is surely not an exact symmetry of our universe.

Assuming that the standard model is correct up to the Planck scale or some 10^{18} GeV, the early universe cosmology of the Higgs field poses an important challenge. Given the Higgs mass that has been measured at 125 GeV and the current measurement of the top quark mass, the electroweak vacuum is at best metastable [2]. In fact, the Higgs quartic coupling, which governs the form of the Higgs potential, turns negative at an energy scale $\Lambda \sim 10^{10} - 10^{14}$ GeV. Therefore, the electroweak vacuum with the minimum at 246 GeV is not the ground state of the standard model, but there is a lower minimum to the left instead and thus our vacuum is only metastable. This is a problem in an inflationary universe. The consequences of this metastability of the electroweak vacuum for the standard model coupled to an inflationary sector has recently been discussed [3].

In an expanding universe with Hubble constant H, the evolution of the Higgs field h reads

$$\ddot{h} + 3H\dot{h} + \frac{\partial V(h)}{\partial h} = 0 \tag{6.1}$$

where $V(h)$ is the Higgs potential. Even if the Higgs boson starts off at the origin of its potential, it will most likely be excited to higher field values during inflation. In fact, because the Higgs mass is very tiny compared to the inflationary scales, it can considered massless. Quantum fluctuations of the Higgs boson will push it away from the minimum of the potential. Its quantum fluctuations are of order the Hubble constant H. Therefore, for $H > \Lambda$, it is likely that the Higgs will overpass the barrier between the false vacuum in which our universe lives and the lower state true vacuum.

In [3, 4], it was shown that a direct coupling of the Higgs field to the inflaton can crucially change this scenario if this coupling is able to make the Higgs potential convex. This interaction between the inflaton and the Higgs field is able to pull the Higgs boson back to the electroweak vacuum during inflation. This is closely related to an earlier claim [5] that the curvature coupling of the Higgs field can be thought of as an additional mass term $-\xi R$ in the Higgs potential and could thus stabilise the Higgs field. We will argue below that this interpretation of the curvature term is not completely correct, and indeed, the two mechanisms are closely related when carried out correctly. Assuming that there is no new physics between the weak scale and the scale of inflation, we will obtain a new prediction for the value of the non-minimal coupling of the Higgs field to the curvature.

6.2 Non-minimal Couplings and the Mixing of Degrees of Freedom

Before the discovery of the Higgs, cosmologists had already been looking at the non-minimal coupling of scalar fields to the Ricci scalar. In inflationary cosmology, one often works with actions of the following type

$$S_{scalar} = \int d^4x \sqrt{-g} \left(\frac{1}{2} \partial_\mu \phi \partial^\mu \phi - \frac{1}{2} m^2 \phi^2 + \frac{1}{2} \xi \phi^2 R \right), \tag{6.2}$$

where m is the mass of the scalar field ϕ. This coupling has been largely investigated, see e.g. [6–10]. When the Higgs field was found, it became clear that this coupling was not only a toy model that could be implemented in curved space-time, but that it could be phenomenologically relevant.

Before deducing our prediction for the value of the non-minimal coupling of the Higgs field to Ricci scalar, we shall address a frequent misunderstanding which can be very important when studying Higgs physics within the context of cosmology and very early universe physics. It is often claimed that the non-minimal coupling that shows up in Eq. (6.2) of a scalar field to the Ricci scalar is equivalent to a curvature-dependent contribution to the mass of the scalar field. We shall argue that this interpretation is not strictly right. We then show that the non-minimal coupling of the Higgs field to the Ricci scalar does actually help to stabilize the Higgs potential, and moreover, it can even pull the Higgs field back to the false vacuum from a Planckian initial value.

We now address the problem of the Higgs mass. If we naively vary the action for a scalar field ϕ containing the non-minimal coupling (6.2), we get the field equation

$$(\Box + m^2 - \xi R)\phi = 0, \tag{6.3}$$

and it is often claimed that the term ξR is a curvature-dependent mass term for the scalar field ϕ. In an FLRW universe, the Ricci scalar drops from $R = 12H^2$ during inflation, with constant Hubble rate H, to $R \approx 0$ in a radiation dominated era after inflation, which could lead to a sizable production of the Higgs field after inflation [11]. However, this argument is incomplete. The point is that the non-minimal coupling gives rise to a mixing between the kinetic terms of the scalar field and of the metric. We shall illustrate this point with the standard model of particle physics since this is the only known model that so far has a fundamental scalar field which has actually been seen in the laboratory. Nonetheless, the logic can be applied to any scalar field non-minimally coupled to the Ricci scalar.

Starting with the standard model Lagrangian $\mathcal{L}_{SM}$, we have

$$S = \int d^4x \sqrt{-g} \left[\left(\frac{1}{2} M^2 + \xi \mathcal{H}^\dagger \mathcal{H} \right) R - (D^\mu \mathcal{H})^\dagger (D_\mu \mathcal{H}) - \mathcal{L}_{SM} \right] \tag{6.4}$$

where $\mathcal{H}$ is the SU(2) scalar doublet, we will see that this is not actually the Higgs boson of the standard model. After electroweak symmetry breaking, the scalar boson gains a non-zero vacuum expectation value, $v = 246$ GeV, M and ξ are then fixed by the relation

$$(M^2 + \xi v^2) = M_P^2 . \tag{6.5}$$

The clearest way to see that $\mathcal{H}$ is not actually the Higgs boson is by performing a conformal transformation to the Einstein frame [12–14] $\tilde{g}_{\mu\nu} = \Omega^2 g_{\mu\nu}$, where $\Omega^2 = (M^2 + 2\xi\mathcal{H}^\dagger\mathcal{H})/M_P^2$. The action in the Einstein frame is then given by

$$S = \int d^4x\,\sqrt{-\tilde{g}}\Bigg[\frac{1}{2}M_P^2\tilde{\mathcal{R}} - \frac{3\xi^2}{M_P^2\Omega^4}\partial^\mu(\mathcal{H}^\dagger\mathcal{H})\partial_\mu(\mathcal{H}^\dagger\mathcal{H}) \\ - \frac{1}{\Omega^2}(D^\mu\mathcal{H})^\dagger(D_\mu\mathcal{H}) - \frac{\mathcal{L}_{SM}}{\Omega^4}\Bigg]. \tag{6.6}$$

Expanding around the Higgs vacuum expectation value and specializing to unitary gauge, $\mathcal{H} = \frac{1}{\sqrt{2}}(0, \phi + v)^\top$, we see that in order to have a canonically normalized kinetic term for the physical Higgs field we need to perform the redefinition

$$\frac{d\chi}{d\phi} = \sqrt{\frac{1}{\Omega^2} + \frac{6\xi^2 v^2}{M_P^2\Omega^4}}, \tag{6.7}$$

where χ denotes the physical Higgs. Expanding $1/\Omega$, we see at leading order the field redefinition only has the effect of a wave function renormalization of $\phi = \chi/\sqrt{1+\beta}$ where $\beta = 6\xi^2v^2/M_P^2$. Thus the canonically normalized scalar field, i.e., the true Higgs boson, does not have any special coupling to gravity and it couples like any other field to gravity in agreement with the equivalence principle.

The same conclusion can also be reached in the Jordan frame (6.4). After fully expanding the Higgs field around its vacuum expectation value and also the metric around a fixed background, $g_{\mu\nu} = \bar{\gamma}_{\mu\nu} + h_{\mu\nu}$, we get a term proportional to $\xi v\phi\Box h^\mu_\mu$:

$$\mathcal{L}^{(2)} = -\frac{M^2 + \xi v^2}{8}(h^{\mu\nu}\Box h_{\mu\nu} + 2\partial_\nu h^{\mu\nu}\partial_\rho h^{\mu\rho} - 2\partial_\nu h^{\mu\nu}\partial_\mu h^\rho_\rho - h^\mu_\mu\Box h^\nu_\nu \\ + \frac{1}{2}(\partial_\mu\phi)^2 + \xi v(\Box h^\mu_\mu - \partial_\mu\partial_\nu h^{\mu\nu})\phi \tag{6.8}$$

After correctly diagonalizing the kinetic terms and canonically normalizing the Higgs field and graviton via

$$\phi = \chi/\sqrt{1+\beta} \tag{6.9}$$

$$h_{\mu\nu} = \frac{1}{M_P}\tilde{h}_{\mu\nu} - \frac{2\xi v}{M_P^2\sqrt{1+\beta}}\bar{\gamma}_{\mu\nu}\chi, \tag{6.10}$$

we find again that the physical Higgs boson gets renormalized by a factor $1/\sqrt{1+\beta}$.

These findings show that the non-minimal coupling does not lead to stronger gravitational interactions for the Higgs field once it has been properly canonically normalized. We emphasize that the underlying reason is that there is no violation of the equivalence principle. Our results are in sharp contrast to the claims made in [15]. The only known bound to date on the non-minimal coupling of the Higgs field to Ricci scaler continues to be the one obtained in [1], namely that its non-minimal coupling is smaller than 2.6×10^{15}. Although the fact that we might be living in a metastable state is troublesome for the Higgs field in an inflationary scenario, the non-minimal coupling of the Higgs to Ricci scalar does not introduce a new issue. On the contrary, we will now prove that this non-minimal coupling could actually solve the stability problem.

6.3 Vacuum Stability in an Inflationary Universe

We shall now investigate the coupling of the Higgs particle to an inflaton potential $V_I(\sigma)$ that shows up due to the transformation from the Jordan frame to the Einstein frame. As a matter of fact, even if no direct coupling of the Higgs field to the inflaton is assumed in the Jordan frame, it will necessarily appear in the Einstein frame:

$$V_I(\sigma) \rightarrow \frac{V_I(\sigma\Omega)}{\Omega^4} = \frac{V_I(\sigma\Omega)}{\left(1 + \frac{2\xi v\phi(\chi)+\xi\phi(\chi)^2}{M_P^2}\right)^2}, \tag{6.11}$$

but keep in mind that the inflaton σ is not canonically normalized.

Let us first examine the Higgs field values $v \ll \phi \ll M_P|\xi|^{-1/2}$. In this case, we instantly observe that

$$\frac{V_I(\sigma\Omega)}{\Omega^4} \approx V_I(\sigma)\left(1 - 2\xi\phi^2/M_p^2\right). \tag{6.12}$$

A coupling of the inflaton to the Higgs boson is created as a consequence of the mapping to the Einstein frame. Observe that there is a priori no reason to keep out a coupling of the kind $V_I\mathcal{H}^\dagger\mathcal{H}$ in the Jordan frame where the theory is set up. There could be cancellations between this coupling and that created by the transformation to the Einstein frame. Therefore, the value of the coupling of the Higgs field to the inflaton that appears in the Einstein frame cannot be seen as a prediction of the model. We shall ignore a direct inflaton-Higgs coupling in the Jordan frame for the time being and proceed with our study of the induced coupling. We shall now show that a non-minimal coupling of the Higgs field to Ricci scalar can improve some of the issues associated with Higgs cosmology without requiring physics beyond the standard model of particle physics.

In the early universe, we need to consider large Higgs field values ($\phi \gg v$). As explained before, even if one is willing to fine-tune the initial condition for the value of the Higgs field, it will face quantum fluctuations of the order of the Hubble scale H. Unless the Hubble rate is much smaller than the energy scale at which the electroweak vacuum becomes unstable, the Higgs boson is likely to go over the potential barrier, ending up at the lower true vacuum of the theory. We will show, however, that a non-minimal coupling of the Higgs to the curvature could actually solve this issue by generating a direct coupling of the Higgs field to the inflaton should the Jordan frame action contain an inflationary potential V_I.

It has been shown that a direct coupling of the Higgs field to the inflaton can pull the Higgs field [3] back to the false electroweak vacuum quickly during inflation even if the Higgs initial value is chosen to be large. There are essentially three scenarios for the beginning of inflation: the thermal initial state [16], ab initio creation [17, 18] and the chaotic initial state [19, 20]. The thermal initial state starts from a temperature just below the Planck scale, which introduces thermal corrections to the Higgs potential, preventing vacuum decay until the temperature falls to the inflationary de Sitter temperature, at which point it becomes a question of vacuum fluctuation as to whether the Higgs survives in the false vacuum. Nonetheless, the consistency of the thermal equilibrium of the standard model fields when the Higgs assumes a large value has not yet been verified. The ab initio creation is an attractive alternative, where the Higgs would nucleate at the top of the potential barrier. In this case also, stability depends on the size of vacuum fluctuations during inflation. The final possibility, the chaotic initial state, would have the Higgs field starting out at arbitrarily high values. The most likely initial values would be larger than the instability scale Λ, keeping the Higgs field from entering the false vacuum. An anthropic argument could be used to exclude these initial conditions, but we will see that the non-minimal curvature coupling of the Higgs boson can force the Higgs into the false vacuum without anthropic considerations.

As we have already seen, the Einstein frame potential is given by

$$V_E = \frac{V_I(\sigma) + V_\phi(\phi)}{(1 + \xi\kappa^2\phi^2)^2} \tag{6.13}$$

where $\kappa^2 = 8\pi G$. Note that V_E is the total potential in the Einstein frame and it accounts for both the inflaton potential in (6.11) and the Higgs potential. The inflationary expansion rate H_I is the expansion rate of the false vacuum,

$$H_I^2 = \frac{V_I(\sigma)}{3M_p^2}. \tag{6.14}$$

The most severe chaotic initial condition, and the one relevant to eternal chaotic inflation, is the one where V_E is near the Planck scale. For an unstable Higgs potential V_ϕ, $V_E \sim M_p^4$ is only admissible when $\xi < 0$, as illustrated in Fig. 6.1.

Let us indicate by ϕ_m the value of the field at which the potential becomes zero,

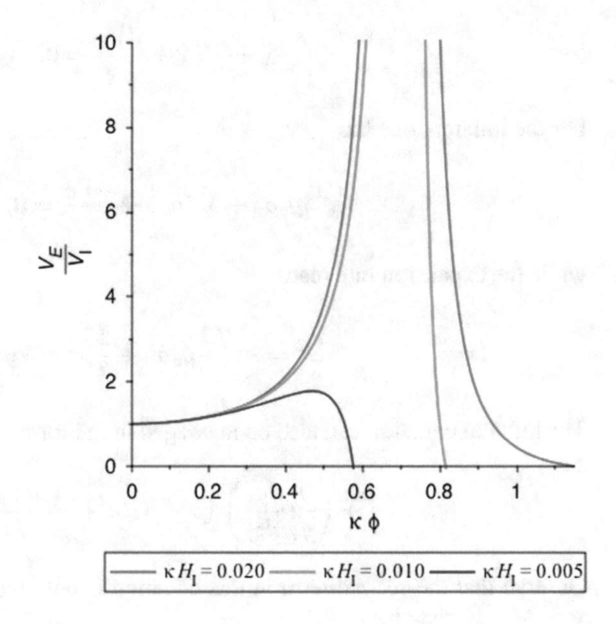

Fig. 6.1 The Einstein frame Higgs potential $V_E(\phi)$ for different values of the false-vacuum inflation rate H_I for $\xi = -2$. The potential vanishes at $\phi = \phi_m$, and there is an asymptote at $\phi = \phi_c$. Consistency of the model (no ghosts) requires $\phi < \phi_c$. An initial condition $V_E \sim M_p^4$ can be achieved with the initial ϕ close to ϕ_c

$$V_I(\sigma) + V_\phi(\phi_m) = 0. \tag{6.15}$$

Observe that ϕ_m depends on H_I. The asymptote in the potential is at ϕ_c,

$$1 + \xi \phi_c^2 / M_p^2 = 0. \tag{6.16}$$

Given that $\phi_c < \phi_m$, then there exists an initial value of ϕ close to ϕ_m at which $V_E \sim M_p^4$ (observe that it was shown in [21] that even with a large non-minimal coupling of the Higgs boson to the Ricci scalar, the cutoff of the effective field theory can be as large as the Planck scale), since $\phi = \phi_c$ is an asymptote. If $\phi_c > \phi_m$, then there is no such value.

Beginning from the initial value, the Higgs evolves to small field values on a timescale comparable to the Hubble expansion rate. Unfortunately, we cannot merely expand the conformal factor in the denominator of the Einstein frame potential for all values of ξ. Nonetheless, it is easy to observe this effect from kinetic terms of the Higgs field and of the inflaton. The kinetic terms for the Higgs field and inflaton are multiplied by g_ϕ and g_σ respectively, where

$$g_\phi = \frac{1 + \xi \kappa^2 \phi^2 + 6\xi^2 \kappa^2 \phi^2}{(1 + \xi \kappa^2 \phi^2)^2}, \qquad g_\sigma = \frac{1}{(1 + \xi \kappa^2 \phi^2)^2} \tag{6.17}$$

Observe that, while one is able to canonically normalise the Higgs field χ as we had done previously, it is not possible to canonically normalise both the Higgs and inflaton fields simultaneously because the field space metric is curved.

The early evolution of the Higgs field is given by the equation

$$\ddot{\chi} + 3H\dot{\chi} + \frac{dV_E}{d\chi} = 0. \tag{6.18}$$

For the inflaton, one has

$$(g_\sigma \dot{\sigma})^{\cdot} + 3H g_\sigma \dot{\sigma} + \frac{dV_E}{d\sigma} = 0, \tag{6.19}$$

while the expansion rate reads

$$3H^2 = \kappa^2 \left(\frac{1}{2} g_\sigma \dot{\sigma}^2 + \frac{1}{2} \dot{\chi}^2 + V_E \right). \tag{6.20}$$

The inflaton equation can also be massaged in the form

$$\ddot{\sigma} + \left(\frac{1}{g_\sigma} \frac{dg_\sigma}{d\chi} \right) \dot{\chi} \dot{\sigma} + 3H\dot{\sigma} + \frac{1}{g_\sigma} \frac{dV_E}{d\sigma} = 0. \tag{6.21}$$

Observe that the second term in this equation is not taken into account in [3]. For $\chi > M_p$, we then have

$$V_E \approx (V_I + V_\phi) e^{\sqrt{8/3}\kappa(\chi - \chi_0)}, \qquad g_\sigma \approx e^{\sqrt{8/3}\kappa(\chi - \chi_0)}. \tag{6.22}$$

There is thus a quick evolution of χ and slow evolution of σ (assuming slow-roll conditions on V_I). In fact, the inflaton evolves on a longer timescale than the Higgs field, leaving a gradual reduction in H_I, and also ϕ_m. Eventually, the potential evolves to $\phi_c > \phi_m$, but at all stages the Higgs field remains in the false vacuum side of the potential barrier. As long as the vacuum fluctuations do not cause quantum tunnelling, the Higgs boson will go in the false vacuum.

The condition that $\phi_c < \phi_m$ implies bounds on the curvature coupling ξ. In order to find these bounds we need to compute ϕ_m from (6.15), and this necessitates an expression for the Higgs potential. For a standard model Higgs field, the large field Higgs potential in flat space reads

$$V_\phi = \frac{1}{4} \lambda(\phi) \phi^4 \tag{6.23}$$

In curved space, the Higgs gets a mass of order H multiplied by Higgs couplings, but we can think of this as a radiative correction to ξ and regard ξ as the effective curvature coupling at the inflationary scale. Other curvature corrections to the Higgs potential might well be significant, but for now these will be neglected.

The effective Higgs coupling $\lambda(\phi)$ goes to zero at some high value of ϕ which we identify as the instability scale Λ. The value of Λ is very strongly dependent on the top quark mass, and currently all we can say is that it lies in the range $10^9 - 10^{18}$ GeV. Moreover, including additional particles to the standard model changes the instability scale (or removes the instability altogether). It is therefore appropriate

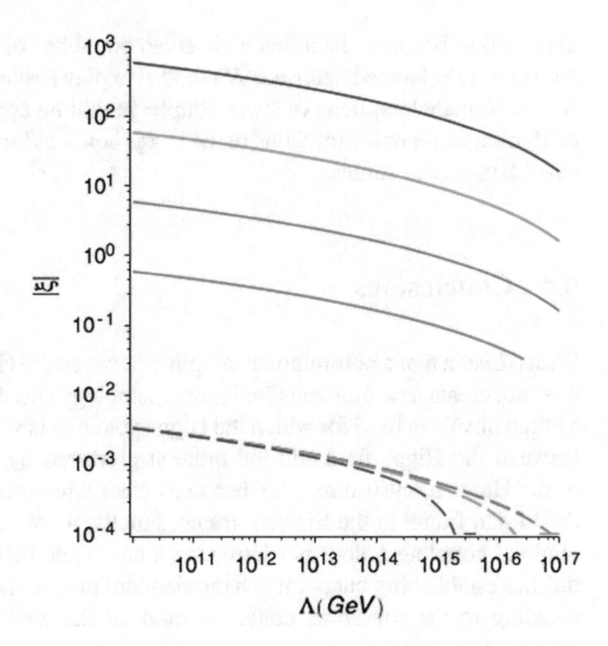

Fig. 6.2 The lower bound on $-\xi$, where ξ is the curvature coupling, for consistent chaotic initial conditions on the Higgs field which will lead the Higgs into the false vacuum. The horizontal axis is the Higgs stability scale. The different curves from bottom to top are for the false vacuum Hubble parameter $0.1M_p$ to $10^{-4}M_p$. The dashed lines show the lower bound for quantum stability of the false vacuum

to provide results treating Λ as a free parameter. In the range of Higgs field values where the potential barrier belongs, we use an approximation to the running coupling given by

$$\lambda(\phi) \approx b\left\{\left(\ln\frac{\phi}{M_p}\right)^4 - \left(\ln\frac{\Lambda}{M_p}\right)^4\right\}, \tag{6.24}$$

with $b \approx 0.75 \times 10^{-7}$. This fits quite well to the renormalisation group calculations [2].

The plots in Fig. (6.2) illustrate numerical results for the values of $-\xi$ which are lower bounds of the range which is consistent with chaotic initial conditions. Also shown by the dashed lines are the quantum bounds from the vacuum tunnelling rate $\exp(-8\pi^2\Delta V_E/3H_I^2) \sim O(1)$, where ΔV_E is the height of the potential barrier [18]. (The quantum bound on $-\xi$ is lower than the one quoted in [15], which we believe is due to our inclusion of the $8\pi^2/3$ factor.) The results show curves for different values of the false vacuum Hubble parameter, basically associated to different initial values of the inflaton field through (6.14). We anticipate that this initial Hubble parameter is near the Planck scale. As advertised, a non-minimal coupling between the Higgs field and the Ricci scalar can pull the Higgs back to the false vacuum of the standard model early on during inflation. Instead of being a source of issues, it can solve some of the problems corresponding to the cosmological evolution of the Higgs boson.

It is worth saying as well that our results also imply that the non-minimal coupling of the Higgs boson to the curvature will not influence reheating as long as the Higgs field value is small during inflation. Reheating could be produced by a direct coupling of the Higgs boson to the inflaton via either couplings of the type $\sigma^2\mathcal{H}^\dagger\mathcal{H}$ or $\sigma\mathcal{H}^\dagger\mathcal{H}$. As usual right-handed neutrinos N could also play a role in reheating via a coupling $\bar{N}N\sigma$. Nonetheless, none of these couplings will be appreciably influenced by the conformal factor or the rescaling of the Higgs boson as long as one is only considering small Higgs field values.

6.4 Conclusions

We have seen that a non-minimal coupling between the Higgs field and the curvature does not create new problems for Higgs cosmology and that, on the contrary, there is a range of values for ξ for which the Higgs potential is stabilized due to the coupling between the Higgs field and the inflaton produced by the non-minimal coupling of the Higgs to curvature. This becomes clear when transforming the theory from the Jordan frame to the Einstein frame. Finally, it was shown in [21] that the non-minimal coupling ξ does not introduce a new scale below the Planck mass which finishes establishing our point that the standard model, should we add a non-minimal coupling to the curvature, could be valid all the way up the Planck scale in an inflationary universe.

References

1. Atkins M, Calmet X (2013) Bounds on the nonminimal coupling of the Higgs boson to gravity. Phys Rev Lett 110(5):051301
2. Degrassi G, Di Vita S, Elias-Miro J, Espinosa JR, Giudice GF, Isidori G, Strumia A (2012) Higgs mass and vacuum stability in the Standard Model at NNLO. JHEP 08:098
3. Lebedev O, Westphal A (2013) Metastable electroweak vacuum: implications for inflation. Phys Lett B 719:415–418
4. Lebedev O (2012) On stability of the electroweak vacuum and the Higgs portal. Eur Phys J C 72:2058
5. Espinosa JR, Giudice GF, Riotto A (2008) Cosmological implications of the Higgs mass measurement. JCAP 0805:002
6. Chernikov NA, Tagirov EA (1968) Quantum theory of scalar fields in de Sitter space-time. Ann Inst H Poincare Phys Theor A9:109
7. Callan CG Jr, Coleman SR, Jackiw R (1970) A new improved energy-momentum tensor. Ann Phys 59:42–73
8. Frommert H, Schoor H, Dehnen H (1999) The cosmological background in the Higgs scalar-tensor theory without Higgs particles. Int J Theor Phys 38:725–735
9. Cervantes-Cota JL, Dehnen H (1995) Induced gravity inflation in the standard model of particle physics. Nucl Phys B 442:391–412
10. Bezrukov FL, Shaposhnikov M (2008) The standard model Higgs boson as the inflaton. Phys Lett B 659:703–706
11. Herranen M, Markkanen T, Nurmi S, Rajantie A (2015) Spacetime curvature and Higgs stability after inflation. Phys Rev Lett 115:241301
12. van der Bij JJ (1994) Can gravity make the Higgs particle decouple? Acta Phys Polon B25:827–832

13. Zee A (1979) A broken symmetric theory of gravity. Phys Rev Lett 42:417
14. Minkowski P (1977) On the spontaneous origin of Newton's constant. Phys Lett 71B:419–421
15. Herranen M, Markkanen T, Nurmi S, Rajantie A (2014) Spacetime curvature and the Higgs stability during inflation. Phys Rev Lett 113(21):211102
16. Guth AH (1981) The inflationary universe: a possible solution to the horizon and flatness problems. Phys Rev D 23:347–356
17. Vilenkin A (1983) The birth of inflationary universes. Phys Rev D 27:2848
18. Hawking SW, Moss IG (1982) Supercooled phase transitions in the very early universe. Phys Lett 110B:35–38
19. Linde AD (1983) Chaotic inflation. Phys Lett 129B:177–181
20. Linde AD (1986) Eternally existing selfreproducing chaotic inflationary universe. Phys Lett B 175:395–400
21. Calmet X, Casadio R (2014) Self-healing of unitarity in Higgs inflation. Phys Lett B 734:17–20

Chapter 7
Conclusions

In this thesis, we have studied classical and quantum extensions of general relativity and their applications to gravitational waves, inflation and dark matter. We focused on effective field theories as they arise in the low energy limit of any UV completion, thus allowing one to investigate gravitational phenomena in a model-independent way.

In Chap. 2, we have shown that modifying the gravitational sector is not really different from modifying the matter sector. One unavoidably includes new degrees of freedom when the Einstein-Hilbert action is complemented with higher curvature invariants. Whether these new degrees of freedom belong to the matter or gravitational sector is just a matter of interpretation, thus not affecting the observables. We then used this equivalence to argue that dark matter could equally be described by a modified gravity model. It is important to stress that, by the time of writing, there is no generally accepted theory that explains the anomalous galaxy rotation curves. Nonetheless, whatever the theory for dark matter that turns out to be right, there will always exist a modified gravity equivalent version of it.

Then in Chap. 3, we studied gravitational waves using the effective field theory approach to quantum gravity. As argued in Chap. 2, modifications of general relativity inevitably leads to new degrees of freedom. In quantum gravity, this is no different. We showed that new degrees of freedom appear in the form of complex poles in the dressed propagator of the graviton, i.e. the propagator containing one-loop quantum corrections. These new poles contribute to new modes of oscillation of gravitational waves and, because they are complex, they lead to a damping in gravitational waves. The damping forces the wave to lose energy to the environment, so it becomes crucial to take this effect into account when inferring the energy released during the merger of black holes. From the bound on the graviton mass found by LIGO, we could constrain the number of fields present in a fundamental theory of gravity.

In Chap. 4, we extended the study of gravitational waves and calculate the energy carried away by the complex modes. By employing the short-wave formalism, we were able to calculate the energy-momentum tensor of gravitational

© Springer Nature Switzerland AG 2019

I. Kuntz, *Gravitational Theories Beyond General Relativity*,
Springer Theses, https://doi.org/10.1007/978-3-030-21197-4_7

waves in quantum gravity. The energy density then follows directly from the energy-momentum tensor as usual. In addition to the term due to classical general relativity, another term that depends on modifications of the dispersion relation shows up. A direct comparison with the expression for the energy density with LIGO's data permits us to find the first constraint on the amplitude of the complex mode. We also showed how the gravitational wave equation in a flat spacetime can be generalized in a curved spacetime by a simple "minimal coupling" prescription.

In Chap. 5 we started the study of inflation via a new model proposed in [1] which combines ideas from Higgs and Starobinsky inflation. We showed that Starobinsky gravity can naturally show up in the formalism of effective field theory. In fact, the square of the Ricci scalar is required for renormalization purposes. In addition, we showed that the coefficient of R^2 flows to the required value in the Starobinsky model when the coefficient of the non-minimal coupling between the Higgs boson and gravity is large. Hence, the Higgs boson is able to trigger Starobinsky inflation via its coupling to gravity. This avoids instability issues caused by large values of the Higgs boson as the scalaron in the Starobinsky model is the only field required to take large values in the early universe.

We continued the study of inflation in Chap. 6 through the non-minimal coupling of the Higgs boson to gravity. We showed that, after diagonalizing and canonically normalizing the action, the induced coupling between the inflationary potential and the Higgs is able to rapidly bring the Higgs field back to the false vacuum even when the scale of its fluctuation is higher than the potential barrier. Thus, the induced coupling between the Higgs and the inflaton's potential is able to stabilize the electroweak vacuum. We also considered the problem of quantum tunnelling that can happen between the false and true vacuum of the theory and we established bounds on the coefficient of the coupling between the Higgs and the curvature so that the Higgs boson remains in the electroweak vacuum.

Although this thesis has presented an important step forward in the field of modified gravity, many problems remain unaddressed. Particularly, there is still a plethora of models seeking elucidation of the dark sector, of inflation and of quantum gravity. In order to rule out some of them, more accurate data are necessary. Upcoming data from LIGO, LISA, Planck and other collaborations should help us on this matter. But in the meantime, while we wait for higher precision experiments, we should concentrate our efforts in theoretical and phenomenological aspects of the effective field theory of gravity as they are model-independent and, in principle, should correctly describe gravity all the way up to the Planck scale. Clearly, at the Planck scale the effective field theory breaks down and one must start worrying about possible UV completions. This is the greatest limitation of the formalism presented in this thesis as we cannot use it to study super-Planckian phenomena. In addition, the effective field theory approach does not address certain conceptual problems in quantum gravity, such as the problem of time. Nonetheless, it provides a systematic way of calculating observables and making falsifiable predictions.

We finish this thesis by indicating potential research directions:

- Can the effective field theory of gravity solve the problem of singularities? It is generally accepted that a quantum theory of gravity should be able to get rid of the singularities of general relativity. While we are still far from finding the UV completed theory for quantum gravity, quantum gravitational effects in the infrared could shed some new light on the problem.
- Black holes are known to cast shadows in their surroundings that are formed due to an extreme type of light bending, forcing photons to get in orbit around them. These shadows carry important information about the spacetime and have distinct phenomenological signatures that can be used to probe the differences among modified gravity theories and further constrain effective theories of gravity.
- General relativity is known to be plagued with pathologies such as traversable wormholes and closed timelike curves. If these solutions were real, they would allow for time travel backwards in time, which would violate causality. Can quantum gravity in the infrared rule out these possibilities?
- A natural extension of the formalism used to calculate the one-loop effective action of quantum gravity would be to consider the Palatini procedure, where the metric and the connection are seen as independent variables. In classical general relativity, varying with respect to the metric and to the connection separately still produces Einstein's equations. However, when quantum corrections are taken into account, this equivalence between the metric and Palatini formalisms no longer holds. The latter could lead to new insights on quantum gravity.

Reference

1. Calmet X, Kuntz I (2016) Higgs Starobinsky inflation. Eur Phys J C **76**(5), 289 (2016)

Appendix A
Perturbative Unitarity

It has been shown in [1] that a large non-minimal coupling of the Higgs to the Ricci scalar does not lead to a new physical scale. While perturbative unitarity appears to be naively violated at an energy scale of M_P/ξ, it can be shown by resumming an infinite series of one-loop diagrams in the large ξ and large N limits but keeping $\xi G_N N$ small that perturbative unitarity is restored (this phenomenon has been called self-healing by Donoghue). In this limit one finds

$$iD^{\alpha\beta\mu\nu}_{dressed} = -\frac{i}{2s}\frac{L^{\alpha\beta}L^{\mu\nu}}{\left(1-\frac{sF_1(s)}{2}\right)}. \tag{A.1}$$

where $L^{\alpha\beta} = \eta^{\alpha\beta} - q^\alpha q^\beta/q^2$, $s = q^2$ and

$$F_1(q^2) = -\frac{1}{30\pi}N_s G_N(\bar{h})(1+10\xi+30\xi^2)\log\left(\frac{-q^2}{\mu^2}\right). \tag{A.2}$$

The background dependent Newton's constant is given by

$$G_N(\bar{h}) = \frac{1}{8\pi(M^2+\xi\bar{h}^2)}. \tag{A.3}$$

In the model described in Chap. 5, one has $\bar{h} = v$. Note that $F_1(s)$ is negative, there is thus no physical pole in the propagator. The dressed amplitude in the large ξ and large N limits is given by

$$A_{dressed} = \frac{48\pi G_N(\bar{h})s\xi^2}{1+\frac{2}{\pi}G_N(\bar{h})s\xi^2\log(-s/\mu^2)} \tag{A.4}$$

© Springer Nature Switzerland AG 2019

I. Kuntz, *Gravitational Theories Beyond General Relativity*,
Springer Theses, https://doi.org/10.1007/978-3-030-21197-4

One easily verifies that the dressed amplitude of the partial-wave with angular momentum $J = 0$ fulfills

$$|a_0|^2 = \mathrm{Im}\,(a_0)\,, \tag{A.5}$$

where a_0 is the amplitude of the $J = 0$ partial wave. In other words, unitarity is restored within general relativity without any new physics or strong dynamics (we are keeping ξG_N small) and there is no new scale associated with the non-minimal coupling despite naive expectations. The cut-off of the effective theory is thus the usual Planck scale.

Reference

1. Calmet X, Casadio R (2014) Self-healing of unitarity in Higgs inflation. Phys Lett B 734:17–20

Printed in the United States
By Bookmasters